# Fundamentals of Pneumatics and Hydraulics

Md. Abdus Salam

# Fundamentals of Pneumatics and Hydraulics

 Springer

Md. Abdus Salam
School of Workforce Development,
Continuing Education and Online Learning
Conestoga College Institute of Technology
and Advanced Learning
Kitchener, ON, Canada

ISBN 978-981-19-0857-6        ISBN 978-981-19-0855-2   (eBook)
https://doi.org/10.1007/978-981-19-0855-2

This Springer imprint is published by the registered company Springer Nature Singapore Pte Ltd.
The registered company address is: 152 Beach Road, #21-01/04 Gateway East, Singapore 189721, Singapore

*I would like to dedicate this book to my respected mother Mst Halima Khatun.*

*Md. Abdus Salam*

# Preface

Most of the Institutions were offering conventional engineering programs such as mechanical, petroleum, civil, chemical, electrical, and computer engineering. Graduates with distinct characteristics are in high demand in the manufacturing industry. Modern programs such as engineering degrees with mechatronics, robotics, energy management, automation, electrical motion, and control management can produce graduates with distinct characteristics. Conestoga College is one of Canada's leading institutions for these types of programs. Other universities and colleges have prioritized revising their present curriculum to include certain new programs in response to industry demand.

Fundamentals of Pneumatics and Hydraulics is a textbook that covers all key topics that are required in the manufacturing industry. The main concepts in this book are given in an easy-to-understand manner for students in technical colleges and universities. The majority of fluid power theories have been addressed using relevant mathematical equations, with some numerical examples and practice problems added to enhance further students learning.

This book is aimed at two main audiences: students and researchers who are studying 'Fundamentals of Pneumatics and Hydraulics' in their academic institutions and believe that their understanding needs to be enhanced further. This book will also provide students from all backgrounds with a better understanding of the fundamentals of fluid power.

Mechatronics engineers, mechanical engineers, and other manufacturing industry specialists make up the second significant audience. Fundamentals of Pneumatics and Hydraulics could potentially benefit this segment of the audience.

I believe that this book can make a meaningful contribution to the students, readers, and relevant system engineers in their practical field.

## Features

The key features of this book are as follows:

- Basic knowledge of direct current and alternating current
- Fundamentals of pneumatics and hydraulics
- Each topic is presented clearly and represented with their mathematical illustration
- Solved examples and practice problems are included after each topic
- Examples are solved by step-by-step methods for easy understanding
- Fluid power system simulation software such as Automation Studio is used for easy understanding
- Large numbers of exercise problems are included at the end of each chapter

## Organization of Books

Basics of electrical terms, Ohms law, Kirchhoff's laws, series–parallel circuits, delta–wye circuits, inductance, capacitance, RL, RC circuits, instantaneous power, apparent power, average power, and reactive power analysis for single-phase and three-phase circuits have been discussed in Chap. 1. Knowledge of alternating current circuits and different types of powers has been discussed in Chap. 3. In Chap. 3, fluid power properties including Reynolds number, continuity equation, Bernoulli's equation, Darcy–Weisbach equation, etc., have been discussed. Basics of hydraulic pump, classification, and power relations have also been included in Chap. 3. In Chap. 4, details of hydraulic pumps have been discussed.

Hydraulic directional valves, classifications, and working principles are discussed in Chap. 5. Hydraulic motors, classification, and efficiency are discussed in Chap. 6. In Chap. 7, different types of hydraulic cylinders and their application have been discussed. Pressure control valves and their operations have been included. Different types of hydraulic pressure control valves and applications are included in Chap. 8. In Chap. 9, the basics of pneumatic and gas laws have been discussed. The pneumatics system components have been discussed in Chap. 10. Electrical control devices, switches, PLC, and applications are discussed in Chap. 11.

## Acknowledgements

I acknowledge Tanya Kell, Dean; Amanda S. Feeser, Chair; Kelly Stedman, Program Manager; Brett Bison, Administrator, Academic Upgrading and Workforce Development, School of Workforce Development, Continuing Education and Online Learning, Conestoga College Institute of Technology, and Advanced Learning, for providing me with a teaching opportunity.

I would also acknowledge my children who helped to read the manuscript.

## Aids for Instructors

Instructors who will adopt this book as a text may obtain the solution manual as a supplement copy by contacting the publishers.

Finally, I believe that this is free of factual errors and omissions. I would like to thank all production staff of Springers who tirelessly work to publish this book successfully.

Kitchener, Canada

Md. Abdus Salam

# Contents

# Chapter 1
# Electrical Parameters and Circuits

## 1.1 Introduction

Electrical components are very important to control the operation of pneumatic and hydraulic cylinders in the fluid power system. Therefore, it is essential to understand the fundamental of electrical terms to comprehend and analyze any electrical applications. This chapter presents these necessary concepts and components as a first step-stone in understanding any underlying electrical principles. In addition, it introduces the core measuring equipment in the electrical domain along with the charge, current, voltage, power, energy, resistance, semiconductor, and insulator.

## 1.2 Charge

The charge is an electrical property of matter, measured in coulombs (C). The charge in motion represents the current and the stationary charge represents static electricity. There are two types of charge namely the positive charge and the negative charge. Like charges repel each other and opposite charges attract each other. The charge is represented by the letter $Q$ or $q$ and its SI-derived unit is the coulomb (C). A coulomb is a large unit of charge and, therefore, generally, the smaller units of charge ($pC$, $nC$, and $\mu C$) are used in practice. Note: charge on an electron, which is negative, has the magnitude of $1.602 \times 10^{-19}$ C. In one Coulomb of charge, there are $1/(1.602 \times 10^{-19}) = 6.24 \times 10^{18}$ electrons [1–3].

## 1.3   Current

The flow of electrons results in the current. The direction of the current is opposite to the flow of electrons (negative charges) as shown in Fig. 1.1. The current is represented by the letter $I$ and its SI unit is Ampere (A) in honour of French Mathematician and Physicist Andre-Marie Ampere (1775–1836).

In general, the current is defined as the total charge $Q$ transferred in time $t$ and it is expressed as,

$$I = \frac{Q}{t} \tag{1.1}$$

The rate of change of charge transferred at any particular time is known as instantaneous current, $i$, and it is expressed as,

$$i = \frac{dq}{dt} \tag{1.2}$$

$$dq = i dt \tag{1.3}$$

$$q = \int_{t_1}^{t_2} i dt \tag{1.4}$$

Equation (1.4) provides the expression for the total charge transferred through a conductor between time $t_1$ and $t_2$.

***Example 1.1***   The expression of charge at a terminal is given by $q = 2t^3 + 5t^2 + t + 1$ C. Find the current at $t = 0.5$ s.

**Solution**

$$i = \frac{dq}{dt} = \frac{d}{dt}(2t^3 + 5t^2 + t + 1) = 6t^2 + 10t + 1 \text{ A} \tag{1.5}$$

At $t = 0$ s, the value of the current is,

**Fig. 1.1**   Direction of current

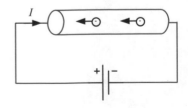

**Fig. 1.2**  Variation of current with time

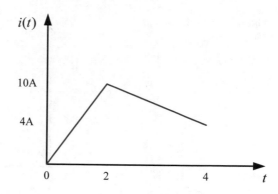

$$i = 6(0.5)^2 + 10(0.5) + 1 = 7.5 \, \text{A} \qquad (1.6)$$

**Practice Problem 1.1**

The total charge entering a terminal is given by $q = 2e^{-2} + 3t^2 + t$ C. Find the current at $t = 0.1$ s.

*Example 1.2*  A current as shown in Fig. 1.2 passes through a wire. Determine the value of the charge.

**Solution**

The expression of current from 0 to 2 s is,

$$i(t) = \frac{10}{2}t = 5t \, \text{A} \qquad (1.7)$$

The expression of current from 2 to 4 s can be written as,

$$\frac{i(t) - 4}{4 - 10} = \frac{t - 4}{4 - 0} \qquad (1.8)$$

$$i(t) - 4 = -\frac{6}{4}(t - 4) \qquad (1.9)$$

$$i(t) = 4 - \frac{6}{4}(t - 4) = 10 - 1.5t \qquad (1.10)$$

The value of the charge from 0 to 4 s can be calculated as,

$$q = \int_0^2 5t\,dt + \int_2^4 (10 - 1.5t)\,dt \qquad (1.11)$$

**Fig. 1.3** Triangular current
wave

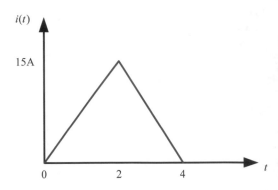

$$q = 2.5(2^2 - 0) + 10(4 - 2) - 0.75(4^2 - 2^2) = 21\,\text{C} \qquad (1.12)$$

**Practice Problem 1.2**
The value of an instantaneous current varying with time is shown in Fig. 1.3. Calculate
the value of the charge.

## 1.4  Direct and Alternating Currents

There are two types of current namely direct current, which is abbreviated as dc and
alternating current, which is abbreviated as ac [4, 5]. When the magnitude of the
current does not change with time, the current is known as direct current, while, in
alternating current, the magnitude of the current changes with respect to the time.
Both of these currents are shown in Fig. 1.4.

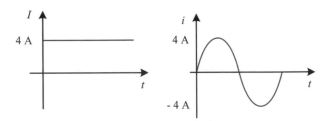

**Fig. 1.4** Direct and alternating current

## 1.5 Conductor, Insulator and Semiconductor

The concept of conductor, insulator and semiconductor can be presented in terms of energy-band models from solid-state physics [6]. The band model consists of two energy bands, valance band and conduction band, and the energy gap between these two bands. The lower-energy valance band contains the charges at their lowest energy levels while the highest-energy conduction band, being generally empty, contains the energy-excited charges to make conduction. In a conductor, a large number of charges populate the conduction band even at room temperature, and as a result, a conductor provides the least resistance to the flow of charges. In this case, there is no energy gap between the valence band and the conduction band, and that means these two energy bands overlap with each other as shown in Fig. 1.5a. For a conductor, the charges in the conduction band are loosely bound with the parent atoms, and hence, they can move easily under the influence of an external energy source such as an electric field.

In the insulator, the energy gap in between the valence and conduction bands is very high, as a result, no charge can gain enough energy to jump to the conduction band as shown in Fig. 1.5b. As a result, an insulator provides the most resistance to the flow of charges.

A semiconductor falls in between the conductor and insulator and it offers moderate resistance to flow the charge. A semiconductor has a smaller energy gap in between the conduction band and the valence band when compared to the insulators, as shown in Fig. 1.5c.

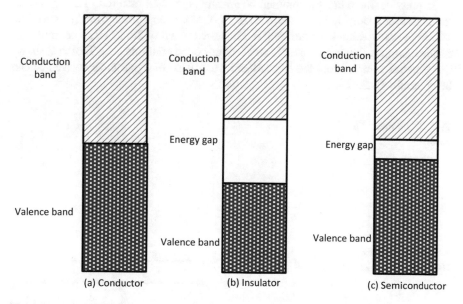

**Fig. 1.5** Energy bands for conductor, insulator and semiconductor

## 1.6   Resistance and Conductance

The property of a material, which opposes the flow of electric current through it, is known as resistance. The resistance is represented by the letter $R$ and its SI unit is Ohm ($\Omega$). The symbol of a resistor is shown in Fig. 1.6. The resistance of any material is directly proportional to the length ($l$), and inversely proportional to the cross-sectional area ($A$) of the material as shown in Fig. 1.7, and can be written as,

$$R \propto l \tag{1.13}$$

$$R \propto \frac{1}{A} \tag{1.14}$$

Combining Eqs. (1.13) and (1.14) yields,

$$R \propto \frac{l}{A} \tag{1.15}$$

$$R = \rho \frac{l}{A} \tag{1.16}$$

where $\rho$ is the proportionality constant which is known as the resistivity of a material and its unit is $\Omega$-m. A good conductor has a very low resistivity, while an insulator has a very high resistivity.

In practice, the common commercial resistors are manufactured in a way so that their resistance values can be determined by the colour bands printed on their surfaces. Different coded values for different colours are shown in Table 1.1. The manufacturing tolerance for each resistor is also colour-coded on its surface; Table 1.2 shows these coded values. From the colour bands, the value of a particular resistance can be determined as,

**Fig. 1.6** Symbol of a resistor

**Fig. 1.7** Conductor with cross-section

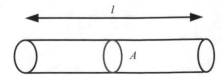

**Table 1.1** Resistance colour codes with values

| Colours name | Colours | Value |
| --- | --- | --- |
| Black | | 0 |
| Brown | | 1 |
| Red | | 2 |
| Orange | | 3 |
| Yellow | | 4 |
| Green | | 5 |
| Blue | | 6 |
| Violet | | 7 |
| Grey | | 8 |
| White | | 9 |

**Table 1.2** Tolerance colour codes with values

| Colours name | Colours | Value (%) |
| --- | --- | --- |
| Orange | | 0.01 |
| Yellow | | 0.001 |
| Brown | | 1 |
| Red | | 2 |
| Gold | | 5 |
| Silver | | 10 |
| No colour | | 20 |

$$R = xy \times 10^z \, \Omega \tag{1.17}$$

*where x is the first colour band, closest to the edge of the resistor, y is the second colour band next to x, and z is the third colour band. The fourth colour band (if any) represents the tolerance in percent.*

Tolerance is the percentage of error in the resistance value that says, how much more or less one can expect a resistor's actual measured resistance to be from its stated resistance.

As an example, the resistance value of a resistor shown in Fig. 1.8, can be determined from its colour code, with the help of Tables 1.1 and 1.2 as,

$$R = 52 \times 10^6 = 52 \, M\Omega \tag{1.18}$$

**Fig. 1.8** Resistance with colours

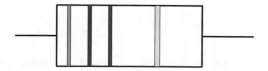

The tolerance is calculated as,

$$Tolerance = 52 \times 10^6 \times 10\% = 5.2 \, M\Omega \tag{1.19}$$

The measured resistance may vary between $(52 + 5.2)$ M$\Omega$ and $(52–5.2)$ M$\Omega$.

Conductance, which is the reciprocal of the resistance, is a characteristic of materials that promotes the flow of electric charge. It is represented by the letter $G$ and its SI unit is Siemens (S). Mathematically, it is expressed as,

$$G = \frac{1}{R} \tag{1.20}$$

The resistance of a good conductor such as copper aluminum etc. increases with an increase in temperature. Whereas the resistance of electrolyte (*An electrolyte is a substance that produces an electrically conducting solution when dissolved in a polar solvent, such as water*), alloy (*An alloy is a metal, made of the combination of two or more metallic elements*) and insulating material decreases with an increase in temperature. Let us assume that the resistance of a conductor at temperature $T$ and $T_0$ degrees centigrade is $R$ and $R_0$, respectively. In this case, the change in resistance is,

$$\Delta R = R - R_0 \tag{1.21}$$

While the change in temperature is,

$$\Delta T = T - T_0 \tag{1.22}$$

The ratio of change in resistance to the resistance at $T_0$ degree centigrade is directly proportional to the change in temperature and it can be expressed as,

$$\frac{R - R_0}{R_0} \propto (T - T_0) \tag{1.23}$$

$$\frac{R - R_0}{R_0} = \alpha(T - T_0) \tag{1.24}$$

$$R - R_0 = \alpha R_0(T - T_0) \tag{1.25}$$

$$R = [R_0 + \alpha R_0(T - T_0)] \tag{1.26}$$

$$R = R_0[1 + \alpha(T - T_0)] \tag{1.27}$$

where $\alpha$ is the proportionality constant which is known as the temperature coefficient of the resistance. Equation (1.27) provides the relationship between the resistance

**Fig. 1.9** Resistance with different temperatures

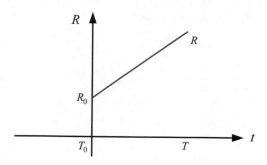

and the temperature, which has been presented in Fig. 1.9. Here, the temperature co-efficient $\alpha$ represents the slope of the line.

***Example 1.3*** The length and diameter of a copper wire are found to be 100 m and 0.002 m, respectively. Consider that the resistivity of the copper wire is $1.72 \times 10^{-8}$ $\Omega$-m. Calculate the value of the resistance.

**Solution**
The area of the wire is,

$$A = \pi d^2 = \pi (0.002)^2 = 1.26 \times 10^{-5} \, \text{m}^2 \tag{1.28}$$

The value of the resistance can be calculated as,

$$R = \rho \frac{l}{A} = 1.72 \times 10^{-8} \times \frac{100}{1.26 \times 10^{-5}} = 0.14 \, \Omega \tag{1.29}$$

**Practice Problem 1.3**
The resistance of a 120 m length aluminum wire is found to be 10 $\Omega$. The resistivity of the aluminum wire is $2.8 \times 10^{-8}$ $\Omega$-m. Determine the diameter of the wire.

## 1.7 Voltage

The voltage is one kind of force required to move a charge in the conductor. Voltage is defined as the work done per unit charge when a charge moves it from one point to another point in a conductor. It is represented by the letter $V$ and its unit is volts (V) in honour of Italian Scientist Alessandro Giuseppe Antonio Anastasio Volta (February 1745–March 1827) who invented the electric battery. Mathematically, the voltage is expressed as,

$$v = \frac{dw}{dq} \tag{1.30}$$

**Fig. 1.10** A tank with water

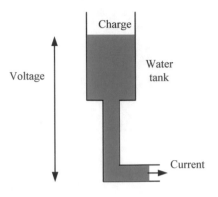

In general, the expression of voltage is,

$$V = \frac{W}{Q} \qquad (1.31)$$

A common analogy, a water tank can be used to describe the charge, voltage, and current. In this analogy, the amount of water in the tank represents the charge, the water pressure represents the voltage and the flow of water through the hosepipe, which is connected at the bottom of the tank, represents the current as shown in Fig. 1.10.

The voltage difference is the difference in potential between the two points in a conductor, which is also known as a potential difference.

**Example 1.4** A charge is required to move between the two points in a conductor. In this case, the expression of work done is given by $w = 3q^2 + q + 1$ J. Find the voltage to move the charge $q = 0.5$ C.

**Solution**
The expression of voltage is,

$$v = \frac{dw}{dq} = \frac{d}{dq}(3q^2 + q + 1) = 6q + 1 \text{ V} \qquad (1.32)$$

The value of the voltage for $q = 0.5$ C is,

$$v = 6q + 1 = 6 \times 0.5 + 1 = 4 \text{ V} \qquad (1.33)$$

**Practice Problem 1.4**
The work done is given by $w = 5q^2 + 2q$ J when a charge is moved between the two points in a conductor. Determine the value of the charge if the voltage is found to be 10 V.

**Fig. 1.11** Ideal voltage and
current sources

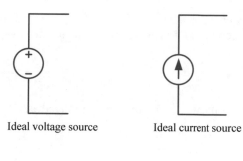

Ideal voltage source          Ideal current source

**Fig. 1.12** Independent
sources

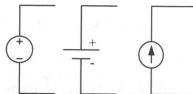

## 1.8 Voltage and Current Sources

Any electrical source (voltage or current) provides energy to the elements connected
to this source. An ideal voltage source is a two-terminal circuit element that provides
a specific magnitude of the voltage across its terminals regardless of the current
flowing through it. An ideal current source is a two-terminal circuit element that
maintains a constant current through its terminals regardless of the voltage across
those terminals. Ideal voltage and current sources are shown in Fig. 1.11.

An active circuit element usually supplies voltage or current to the passive
elements like resistance, inductance and capacitance. An active element is classified as either an independent or dependent (controlled) source of energy. An independent voltage or current source is an active element that can generate a specified
voltage or current without depending on any other circuit elements. The symbols for
independent voltage and current sources are shown in Fig. 1.12.

## 1.9 Electric Power and Energy

Electrical power is defined as the rate of receiving or delivering energy from one
circuit to another circuit. The power is represented by the letter $p$ and its unit is joules
per sec (J/s) or watts (W) in honour of British Scientist James Watt (1736–1819).
Mathematically, the expression of power can be written as,

$$p = \frac{dw}{dt} \tag{1.34}$$

where $p$ is the power in watts (W), $w$ is the energy in joules (J) and $t$ is the times in seconds (s). In general, the total electric power can be written as,

$$P = \frac{W}{t} \qquad (1.35)$$

Equation (1.34) can be re-arranged as,

$$p = \frac{dw}{dq}\frac{dq}{dt} \qquad (1.36)$$

Substituting Eqs. (1.2) and (1.30) into Eq. (1.36) yields,

$$p = vi \qquad (1.37)$$

Equations (1.1) and (1.31) can be re-arranged as,

$$t = \frac{Q}{I} \qquad (1.38)$$

$$W = VQ \qquad (1.39)$$

Substituting Eqs. (1.38) and (1.39) into Eq. (1.35) yields,

$$P = \frac{W}{t} = \frac{VQ}{\frac{Q}{I}} \qquad (1.40)$$

$$P = VI \qquad (1.41)$$

Both Eqs. (1.37) and (1.41) show that the power associated with any basic circuit element is the product of the voltage across that element and the current through that element. If the power, associated with a circuit element, has a positive sign, then the power is absorbed by that element. Whereas, in a circuit element, the power with a negative sign represents that the element delivers power to the other elements. The phenomena of power-delivery and power-absorption by any element can be visualized with the aid of Fig. 1.13.

**Fig. 1.13** Power delivered and absorbed

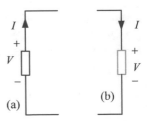

According to the circuit in Fig. 1.18, the expressions of power can be expressed as,

$$P_d = V(-I) = -VI \tag{1.42}$$

$$P_r = VI \tag{1.43}$$

The negative sign in Eq. (1.42) indicates that the power is being delivered by the element, while the positive sign in Eq. (1.43) indicates the power is being absorbed by the element.

Energy is the capacity to do work. Since the SI unit of work is the joule (J), and for the same reason, it is represented by the letter $w$. From Eq. (1.34), the expression of energy can be written as,

$$dw = p\, dt \tag{1.44}$$

For a time-window between $t_0$ and $t$, the total energy can be calculated as,

$$w = \int_{t_0}^{t} p\, dt \tag{1.45}$$

In practice, the energy supplied to the consumers has rated the term of a kilowatt-hour (kWh). One kWh is defined as the electrical power of 1 kW consumed in an hour, and it is expressed as,

$$E = P \times t \tag{1.46}$$

where $E$ is the electrical energy in kWh, $P$ is the power in kW and $t$ is the time in an hour.

**Example 1.5**  The expression of work done is given by $w = 5t^2 + 4t + 1$ J. Calculate the power at $t = 0.1$ s.

**Solution**
The expression of power is,

$$p = \frac{dw}{dt} = \frac{d}{dt}(5t^2 + 4t + 1) = 10t + 4\,\text{W} \tag{1.47}$$

Power at $t = 0.1$ s is,

$$p = 10(0.1) + 4 = 5\,\text{W} \tag{1.48}$$

**Example 1.6** A power with an expression of $p = 5 + 5t^2$ W is used from a time of 0.1 s to 0.4 s to complete a specific task. Calculate the corresponding energy.

**Solution**

$$w = \int\limits_{0.1}^{t=0.4} (5 + 5t^2)\, dt = 5[t]_{0.1}^{0.4} + \frac{5}{3}[t^3]_{0.1}^{0.4}$$

$$= 5(0.4 - 0.1) + 1.67(0.4^3 - 0.1^3) = 1.61\,\text{J} \qquad (1.49)$$

**Example 1.7** A 500 W electric toaster operates for 30 min, a 7 kW DVD player operates for 1 h and a 700 W electric iron operates for 45 min. Find the total energy used, and the associated cost if the energy price is 5 cents per kWh.

**Solution**
The total energy used is,

$$E = \frac{500}{1000} \times \frac{30}{60} + 7 \times 1 + \frac{700}{1000} \times \frac{45}{60}$$
$$= 0.500 \times 0.5 + 7 + 0.700 \times 0.75 = 7.78\,\text{kWh} \qquad (1.50)$$

The cost of the energy used is,

$$C = 7.78 \times \frac{5}{100} = 0.39\,\$ \qquad (1.51)$$

**Practice Problem 1.5**
The power consumed at time $t$, for the work $w = 5t^2 + 4t + 1$ J, is found to be 20 W. Calculate time $t$.

**Practice Problem 1.6**
A power of $p = 10e^{-t} + 2$ W is used to complete a task. Calculate the energy for the time interval 0.01–0.2 s.

**Practice Problem 1.7**
Five 10 W energy-saving lightbulbs operate for 8 h, a 60 W smart TV operates for 5 h and an 800 W hairdryer operates for 30 min. Calculate the total energy used, and the associated cost if the energy price is 20 cents per kWh.

## 1.10  American Wire Gauge

The AWG means American Wire Gauge. It is used as a standard method to find the wire diameter for practical wiring. The AWG is often known as Brown and Sharpe (B&S) wire gauge. Electrical power is defined as the rate of receiving or delivering energy from one circuit to another circuit. The diameter of the wire in inch and a millimeter is calculated as,

$$d_n = 0.005 \times 92^{\frac{36-n}{39}} \text{ in.} \tag{1.52}$$

$$d_n = 0.127 \times 92^{\frac{36-n}{39}} \text{ mm} \tag{1.53}$$

The large diameter increases the area of the conductor and decreases the resistance. The AWG number is calculated as,

$$n = -39 \log_{92}\left(\frac{dn}{0.005 \text{ in.}}\right) + 36 \tag{1.54}$$

$$n = -39 \log_{92}\left(\frac{dn}{0.127 \text{ mm}}\right) + 36 \tag{1.55}$$

The higher the number thinner the wire. The AWG number 12 or 14 is usually used in the household wiring. However, it depends on the load connected in a house. Telephone wire is typically AWG 22, 24 or 26 and the ground wire is usually AWG 6 or 8 or 10.

The circular mil is abbreviated as $CM$. The $CM$ is a unit representing the cross-sectional area of a wire or a cable. The $CM$ for stranded conductor is calculated as,

$$CM = d^2 \times N \tag{1.56}$$

For a single conductor, the $CM$ is calculated as,

$$CM = d^2 \tag{1.57}$$

where $N$ is the number of strands and $d$ is the diameter. The unit conversion 1 in. = 1000 mils is considered to calculate CM. For a diameter of 0.125 in., the $CM$ is calculated as,

$$d = 0.125 \times 1000 = 125 \text{ mils} \tag{1.58}$$

$$CM = d^2 = 125^2 = 15{,}625 \tag{1.59}$$

The square mils ($SM$) of a wire is calculated as,

$$SM = \frac{\pi d^2}{4} = \frac{\pi \times 125^2}{4} = 12{,}271.85 \qquad (1.60)$$

The voltage drop always is there in a closed circuit and it happens due to the resistance of a particular circuit. The general formula for voltage drops in a wire is calculated as,

$$V_d = \frac{M \times K \times I \times L}{CM} \qquad (1.61)$$

where,

$M$ is the phase multiplier $= 2$ for DC and $1\phi$ AC and $= 1.732$ for $3\phi$ AC,
$K$ is the direct current constant $= 12.9\ \Omega$ for copper wire and $= 21.2\ \Omega$ for an aluminum wire
$I$ is current,
$L$ is the length,
$CM$ is the circular mils.

**Example 1.8**  A load carries a current of 25 A from a 120 V, 60 Hz source using a 12 number wire. Calculate the voltage drop if the distance between the load and source is 100 ft. The CM of a 12 number wire is 6528.64.

**Solution**
The voltage drop is calculated as,

$$V_d = \frac{M \times K \times I \times L}{CM} = \frac{2 \times 12.9 \times 25 \times 100}{6528.64} = 9.87\,\text{V} \qquad (1.62)$$

**Practice Problem 1.8**
A load carries a current of 20 A from a 120 V, 60 Hz source using a 14 number wire. Calculate the voltage drop if the distance between the load and source is 80 ft.

## 1.11  Measuring Equipment

Measuring equipment is used to measure the electrical parameters accurately, both analogue, and digital meters are available in practice. In an analogue meter, a pointer moves on a scale to read out the reading. Whereas the digital meter can display the reading directly in the form of numerical digits. Some of these widely used meters have been discussed below.

**Fig. 1.14**  Schematic of an
ohmmeter

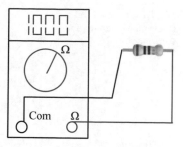

**Fig. 1.15**  Symbol and
connection diagram of an
ammeter

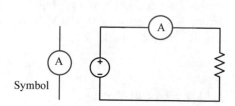

## 1.11.1  Ohmmeter

An ohmmeter, shown in Fig. 1.19, can measure the resistance of an electrical component. To measure the resistance of any component, it needs to be disconnected from
the circuit. An ohmmeter is also used to identify the continuity of a circuit and
to identify the correct terminal-ends in the application field where long cables are
used. In the measurement steps, the meter supplies current to the resistance, then
measures the voltage drop across that resistance. Finally, the meter gives the value
of the resistance (Fig. 1.14).

## 1.11.2  Ammeter

An ammeter is used to measure the current in the circuit. An ammeter is always
connected in series with the circuit under test. This meter offers low internal resistance, which does not affect the actual measurement. The meter symbol and a circuit
under test are shown in Fig. 1.15.

## 1.11.3  Voltmeter

A voltmeter is used to measure the voltage across any circuit element. A voltmeter
is always connected in parallel to the circuit element under test. Voltmeters can be
designed to measure AC or DC.

**Fig. 1.16** Symbol and
connection diagram of a
voltmeter

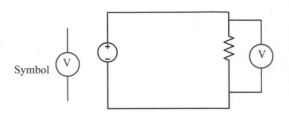

The internal resistance of a voltmeter is usually kept very high to reduce the current flow through it to a negligibly small amount. Most of the voltmeters have several scales including 0–250 V, 0–500 V. The meter symbol and a circuit under test are shown in Fig. 1.16.

### 1.11.4  Wattmeter

A wattmeter is used to measure either the power delivered by an electrical source or the power absorbed by a load. As shown in Fig. 1.17, a wattmeter has two coils, the current coil (CC) and the voltage coil (VC).

The current coil with very low resistance is connected in series with the load and responds to the load current, while the voltage coil is connected in parallel to the load circuit and responds to the load voltage. Depending on the test section (source or load), one terminal of each coil is shorted and connected either to the source or to the load as shown in Fig. 1.18. In an ac circuit, a wattmeter is widely used to measure the single-phase and three-phase real power as shown in Fig. 1.19.

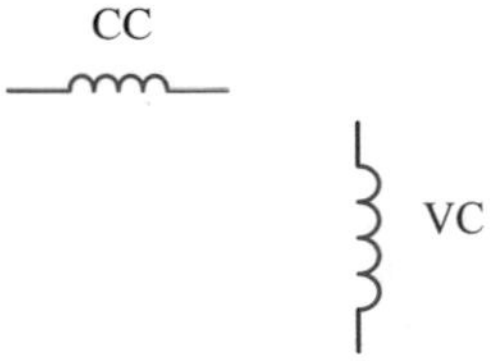

**Fig. 1.17** Current coil and                                              CC
voltage coil of a wattmeter

**Fig. 1.18** Schematic of
wattmeter connection

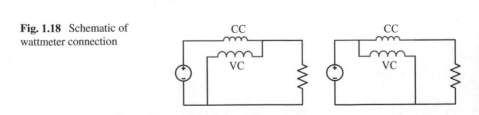

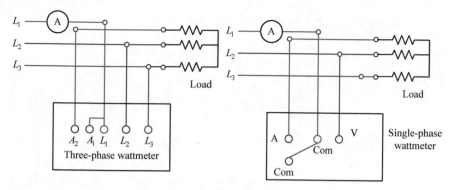

**Fig. 1.19**  Connection diagrams of a single-phase and three-phase of wattmeter

## 1.12  Efficiency

The efficiency of a system is defined as the ratio of output power to input power. The input power is equal to the output power plus the loss as shown in Fig. 1.25. Mathematically, the efficiency $\eta$ is expressed as,

$$\eta = \frac{P_o}{P_{in}} \times 100 \tag{1.63}$$

$$P_{in} = P_o + P_l \tag{1.64}$$

where,

$P_{in}$ is the input power in the system in W,
$P_o$ is the output power in the system in W,
$P_l$ is the loss in the system in W (Fig. 1.20).

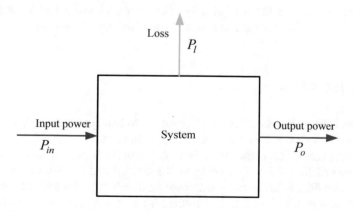

**Fig. 1.20**  A system with different types of powers

Substituting Eq. (1.64) into Eq. (1.63) yields,

$$\eta = \frac{P_{in} - P_l}{P_{in}} \times 100 \tag{1.65}$$

$$\eta = \left(1 - \frac{P_l}{P_{in}}\right) \times 100 \tag{1.66}$$

Again, substituting Eq. (1.64) into Eq. (1.63) and it can be written as,

$$\eta = \frac{P_o}{P_o + P_l} \times 100 \tag{1.67}$$

**Example 1.9** The input power and the loss of a system are found to be 150 W and 30 W, respectively. Determine the output power and the efficiency of the system.

**Solution**
The output of the system can be determined as,

$$150 = P_o + 30 \tag{1.68}$$

$$P_o = 150 - 30 = 120 \, \text{W} \tag{1.69}$$

The efficiency is calculated as,

$$\eta = \frac{P_o}{P_{in}} \times 100 = \frac{120}{150} \times 100 = 80\% \tag{1.70}$$

**Practice Problem 1.9**
The output power and the loss of a system are found to be 100 W and 10 W, respectively. Calculate the input power and the efficiency of the system.

## 1.13  Ohm's Law

Ohm's law is very important in an electric circuit to find out the current, voltage and resistance. Ohm's law states that the current flows in a conductor directly proportional to the voltage across the conductor. A German Physicist, George Simon Ohm (1787–1854) invented the relationship between the current and the voltage for a given resistor. According to his name, it is known as Ohm's law. The current $I$ flows in a resistor $R$ as shown in Fig. 2.1. The voltage across the resistor is $V$. According to Ohm's law, the following relation can be written as,

**Fig. 1.21** Current in a resistor

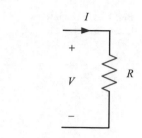

**Fig. 1.22** Variation of voltage with the current

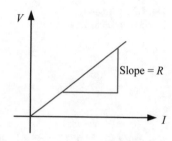

$$I \infty V \tag{1.71}$$

$$V = IR \tag{1.72}$$

where $R$ is the proportionality constant, which is known as a resistance of a circuit (Figs. 1.21 and 1.22).

Equation (1.72) can be expressed as,

$$I = \frac{V}{R} \tag{1.73}$$

In another way, Ohm's law states that the current flows in a resistor is directly proportional to the voltage across the resistor and inversely proportional to the resistance. According to Ohm's law, if increases the current, the voltage will increase as can be seen in Fig. 2.2.

*Example 1.10* A hairdryer is connected to a 220 V source and draws a 3 A current. Determine the value of the resistance.

**Solution**

The value of the resistance can be determined as,

$$R = \frac{V}{I} = \frac{220}{3} = 73.33 \, \Omega \tag{1.74}$$

**Fig. 1.23**  Circuit for
Example 1.11

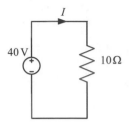

**Fig. 1.24**  Circuit for
Practice Problem 1.11

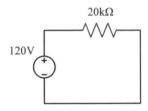

*Example 1.11* A 10 Ω resistance is connected to a voltage source as shown in Fig. 1.23. Calculate the current draws and power absorbed by a resistance.

**Solution**
The current draws by the resistance can be determined as,

$$I = \frac{V}{R} = \frac{40}{10} = 4\,\text{A} \tag{1.75}$$

The power absorbed by the resistance is,

$$P = I^2 R = 4^2 \times 10 = 160\,\text{W} \tag{1.76}$$

**Practice Problem 1.10**
A 600 W blender machine is connected to a 220 V source. Calculate the value of the current and resistance.

**Practice Problem 1.11**
A 20 kΩ resistance is connected to a voltage source as shown in Fig. 1.24. Find the current.

## 1.14   Kirchhoff's Laws

In 1845, German Scientist, Gustav Robert Kirchhoff discovered the relationship between different types of currents and voltages in an electrical circuit. These laws

**Fig. 1.25** Circuit for KCL

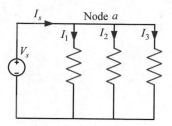

are known as Kirchhoff's laws which are used to calculate the current, and voltage in an electrical circuit.

## 1.14.1 Kirchhoff's Current Law

Kirchhoff's current law states that the algebraic sum of the currents around a node equals zero. Figure 1.25 is used to illustrate Kirchhoff's current law (KCL). Here, the currents entering the node are considered positive and coming out from the node are considered negative.

From Fig. 2.5, the following equation can be written as,

$$I_s - I_1 - I_2 - I_3 = 0 \qquad (1.77)$$

$$I_s = I_1 + I_2 + I_3 \qquad (1.78)$$

$$\sum I_{entering} = \sum I_{leaving} \qquad (1.79)$$

From Eq. (2.9), the KCL can be stated as the sum of the entering currents at a node is equal to the current leaving from the same node.

**Example 1.12** The current distribution in a circuit is shown in Fig. 1.27. Determine the value of the unknown current.

**Solution**
Apply KCL at the node of the circuit in Fig. 1.27 yields, (Fig. 1.26)

$$20 = 12 + I_2 \qquad (1.80)$$

$$I_2 = 20 - 12 = 8\,\text{A} \qquad (1.81)$$

**Fig. 1.26** Circuit for
Example 1.12

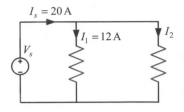

**Fig. 1.27** Circuit for
Practice Problem 1.12

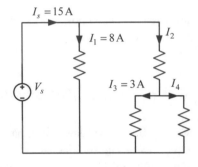

**Practice Problem 1.12**

Figure 1.27 shows a circuit with the current distributions. Calculate the values of the
unknown currents.

### 1.14.2   Kirchhoff's Voltage Law

Kirchhoff's Voltage Law (KVL) is also a basic law that is used in an electrical circuit
to find the unknown voltage, current and resistance. The KVL can be stated as the
sum of the voltage rises and drops around a closed circuit is equal to zero. Figure 1.28
is considered to explain the KVL.

Applying KVL to the circuit in Fig. 1.28 yields,

$$-V_{s1} - V_{s2} + V_1 + V_2 + V_3 = 0 \tag{1.82}$$

**Fig. 1.28** Circuit for KVL

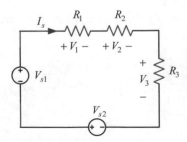

**Fig. 1.29** Circuit for
Example 1.13

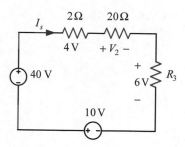

$$V_{s1} + V_{s2} = V_1 + V_2 + V_3 \tag{1.83}$$

$$\sum V_{rises} = \sum V_{drops} \tag{1.84}$$

From Eq. (1.84), the KVL can be re-stated as the sum of the voltage rises is equal to the sum of the voltage drops.

**Example 1.13** The voltage distribution of a circuit is shown in Fig. 1.29. Determine the value of the $V_2$, source current $I_s$, and the resistance $R_3$.

**Solution**
Applying KVL to the circuit in Fig. 1.29 yields,

$$-40 - 10 + 4 + 6 + V_2 = 0 \tag{1.85}$$

$$V_2 = 40 \, V \tag{1.86}$$

The value of the current in the circuit is determined as,

$$V_2 = I_s \times 4 = 40 \tag{1.87}$$

$$I_s = 10 \, A \tag{1.88}$$

The value of the resistance $R_3$ is determined as,

$$R_3 = \frac{V_3}{I_s} = \frac{6}{10} = 0.6 \, \Omega \tag{1.89}$$

**Example 1.14** An electrical circuit with a voltage-controlled voltage source is shown in Fig. 1.30. Find the power absorbed by the 6 $\Omega$ resistance.

**Fig. 1.30** Circuit for
Example 1.14

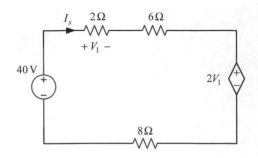

**Solution**
Applying KVL to the circuit in Fig. 1.30 yields,

$$-40 + 2I_s + 6I_s + 8I_s + 2V_1 = 0 \tag{1.90}$$

$$2I_s - V_1 = 0 \tag{1.91}$$

$$V_1 = 2I_s \tag{1.92}$$

Substituting Eq. (1.92) into Eq. (1.90) yields,

$$-40 + 2I_s + 6I_s + 8I_s + 2 \times 2I_s = 0 \tag{1.93}$$

$$I_s = \frac{40}{20} = 2\,\text{A} \tag{1.94}$$

The power absorbed by the 6 $\Omega$ resistance is,

$$P_{6\Omega} = I_s^2 \times 6 = 2^2 \times 6 = 24\,\text{W} \tag{1.95}$$

**Practice Problem 1.13**
Figure 1.31 shows an electrical circuit with different parameters. Calculate the value
of the source current, the voltage across 6 and 12 $\Omega$ resistances.

**Practice Problem 1.14**
An electrical circuit with a voltage-controlled voltage source is shown in Fig. 1.32.
Calculate the voltage across the 12 $\Omega$ resistance.

**Fig. 1.31**  Circuit for
Practice Problem 1.13

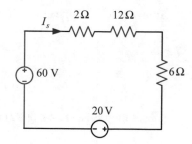

**Fig. 1.32**  Circuit for
Practice Problem 1.14

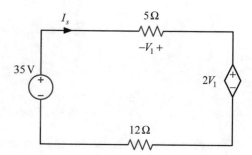

## 1.15  Series Resistors and Voltage Division Rule

In a series circuit, resistances and voltage source are connected end-to-end point,
where the same current flows through each element. Consider three resistances are
connected in series with a voltage source as shown in Fig. 1.33.
　According to Ohm's law, the following equations can be written as,

$$V_1 = I_s R_1 \qquad\qquad (1.96)$$

$$V_2 = I_s R_2 \qquad\qquad (1.97)$$

$$V_3 = I_s R_3 \qquad\qquad (1.98)$$

**Fig. 1.33**  Circuit for voltage
division rule

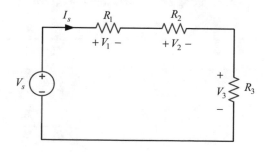

Applying KVL to the circuit in Fig. 1.33 yields,

$$-V_s + V_1 + V_2 + V_3 = 0 \tag{1.99}$$

$$V_s = V_1 + V_2 + V_3 \tag{1.100}$$

Substituting Eqs. (1.96), (1.97) and (1.98) into Eq. (1.100) yields,

$$V_s = I_s R_1 + I_s R_2 + I_s R_3 \tag{1.101}$$

$$V_s = I_s (R_1 + R_2 + R_3) \tag{1.102}$$

$$\frac{V_s}{I_s} = (R_1 + R_2 + R_3) \tag{1.103}$$

$$R_s = (R_1 + R_2 + R_3) \tag{1.104}$$

If a circuit contains $N$ number of series resistances, then the equivalent resistance can be expressed as,

$$R_s = R_1 + R_2 + R_3 + \cdots + R_N = \sum_{n=1}^{N} R_n \tag{1.105}$$

The voltage divider rule or voltage division rule is closely related to the series circuit. From Fig. 1.33, the following equation can be written as,

$$I_s = \frac{V_s}{R_s} \tag{1.106}$$

Substituting Eq. (1.104) into Eq. (1.106) yields,

$$I_s = \frac{V_s}{R_1 + R_2 + R_3} \tag{1.107}$$

Substituting Eq. (1.107) into Eqs. (1.96), (1.97) and (1.98) yields,

$$V_1 = \frac{R_1}{R_1 + R_2 + R_3} V_s \tag{1.108}$$

$$V_2 = \frac{R_2}{R_1 + R_2 + R_3} V_s \tag{1.109}$$

$$V_3 = \frac{R_3}{R_1 + R_2 + R_3} V_s \tag{1.110}$$

**Fig. 1.34** Circuit for
Example 1.15

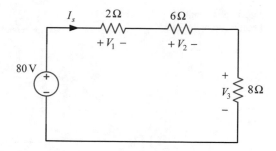

The voltage divider rule for a circuit with $N$ number of resistance is,

$$V_n = \frac{R_n}{R_1 + R_2 + R_3 + \cdots + R_N} V_s \qquad (1.111)$$

From Eq. (1.111), it is concluded that the voltage across any resistance is equal to the total voltage times the ratio of that resistance to the total resistance.

*Example 1.15* A series circuit is shown in Fig. 1.34. Calculate the voltage across each resistance using voltage divider rule.

**Solution**
The voltage across 2 Ω resistance is,

$$V_{2\,\Omega} = \frac{R_1}{R_1 + R_2 + R_3} V_s = \frac{2}{16} \times 80 = 10 \, \text{V} \qquad (1.112)$$

$$V_{6\,\Omega} = \frac{R_2}{R_1 + R_2 + R_3} V_s = \frac{6}{16} \times 80 = 30 \, \text{V} \qquad (1.113)$$

$$V_{8\,\Omega} = \frac{R_3}{R_1 + R_2 + R_3} V_s = \frac{8}{16} \times 80 = 40 \, \text{V} \qquad (1.114)$$

**Practice Problem 1.15**
Figure 1.35 shows a series circuit. Use voltage divider rule to find the source voltage and voltage across 6 Ω resistor.

## 1.16 Parallel Resistors and Current Division Rule

Resistances are sometimes connected in parallel in an electrical circuit. Two resistances are said to be in parallel when they have two common points. In a parallel circuit, the voltage is the same, but currents are different. Consider three resistances are connected in a parallel as shown in Fig. 1.36.

**Fig. 1.35** Circuit for
Practice Problem 1.15

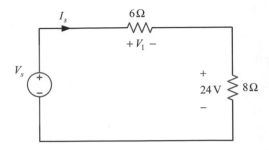

**Fig. 1.36** Three resistors in
parallel

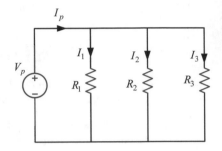

According to Ohm's law, the following equations can be written as,

$$I_1 = \frac{V_p}{R_1} \tag{1.115}$$

$$I_2 = \frac{V_p}{R_2} \tag{1.116}$$

$$I_3 = \frac{V_p}{R_3} \tag{1.117}$$

Applying KCL to the circuit in Fig. 1.36 yields,

$$I_p = I_1 + I_2 + I_3 \tag{1.118}$$

Substituting Eqs. (1.115), (1.116), (1.117) into Eq. (1.118) yields,

$$I_p = \frac{V_p}{R_1} + \frac{V_p}{R_2} + \frac{V_p}{R_3} \tag{1.119}$$

$$I_p = V_p \left( \frac{1}{R_1} + \frac{1}{R_2} + \frac{1}{R_3} \right) \tag{1.120}$$

$$\frac{1}{\left( \frac{1}{R_1} + \frac{1}{R_2} + \frac{1}{R_3} \right)} = \frac{V_p}{I_p} \tag{1.121}$$

$$R_p = \frac{1}{\left(\frac{1}{R_1} + \frac{1}{R_2} + \frac{1}{R_3}\right)} \tag{1.122}$$

If the circuit contains $N$ number of resistances and two resistances, then Eq. (1.122) can be modified as,

$$R_p = \frac{1}{\frac{1}{R_1} + \frac{1}{R_2} + \frac{1}{R_3} + \cdots + \frac{1}{R_N}} \tag{1.123}$$

$$R_p = \frac{1}{\frac{1}{R_1} + \frac{1}{R_2}} \tag{1.124}$$

$$R_p = \frac{1}{\frac{R_1 + R_2}{R_1 R_2}} \tag{1.125}$$

$$R_p = \frac{R_1 R_2}{R_1 + R_2} \tag{1.126}$$

From Eq. (1.126), it is concluded that equivalent resistance for two resistances in parallel is equal to the ratio of the product of two resistances to their sum.

If $R_1 = R_2 = R$, then Eq. (1.126) is modified as,

$$R_p = \frac{R \times R}{2R} = \frac{R}{2} \tag{1.127}$$

From Eq. (1.127), it is concluded that the total or equivalent resistance will be equal to half of the one if two same resistances are in parallel.

The circuit in Fig. 1.37 is considered to explain the current divider rule. From the circuit in Fig. 1.37, the following equations can be written as,

$$R_t = \frac{R_1 R_2}{R_1 + R_2} \tag{1.128}$$

The total voltage across the two resistances are,

$$V_t = I_t R_t \tag{1.129}$$

**Fig. 1.37** Circuit for two resistors in parallel

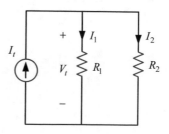

Substituting Eq. (1.128) into Eq. (1.129) yields,

$$V_t = \frac{R_1 R_2}{R_1 + R_2} I_t \tag{1.130}$$

The current in the resistance $R_1$ is,

$$I_1 = \frac{V_t}{R_1} \tag{1.131}$$

Substituting Eq. (1.130) into Eq. (1.131) yields,

$$I_1 = \frac{1}{R_1} \frac{R_1 R_2}{R_1 + R_2} I_t \tag{1.132}$$

$$I_1 = \frac{R_2}{R_1 + R_2} I_t \tag{1.133}$$

The current in the resistance $R_2$ is,

$$I_2 = \frac{V_t}{R_2} \tag{1.134}$$

Substituting Eq. (1.130) into Eq. (1.134) yields,

$$I_2 = \frac{1}{R_2} \frac{R_1 R_2}{R_1 + R_2} I_t \tag{1.135}$$

$$I_2 = \frac{R_1}{R_1 + R_2} I_t \tag{1.136}$$

From Eqs. (1.133) and (1.136), it is concluded that the current through any resistance is equal to the total current times the ratio of the opposite resistance to the sum of the two resistances.

***Example 1.16***  A series–parallel circuit is shown in Fig. 1.38. Calculate the source current and the branch currents.

**Fig. 1.38** Circuit for
Example 1.16

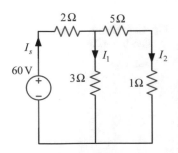

**Fig. 1.39** Circuit for
Example 1.17

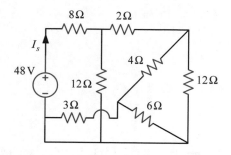

**Solution**
The total resistance of the circuit is calculated as,

$$R_t = 2 + \frac{3 \times (1 + 5)}{3 + 6} = 4\,\Omega \tag{1.137}$$

The value of the source current is calculated as,

$$I_s = \frac{60}{4} = 15\,\text{A} \tag{1.138}$$

The branch currents $I_1$ and $I_2$ can be calculated as,

$$I_1 = \frac{(5 + 1)}{6 + 3} \times 15 = 10\,\text{A} \tag{1.139}$$

$$I_2 = \frac{3}{6 + 3} \times 15 = 5\,\text{A} \tag{1.140}$$

**Example 1.17** Figure 1.39 shows a series–parallel circuit with a jumper wire. Find
the total resistance and the source current.

**Solution**
In this circuit, 3 and 6 $\Omega$ resistances are connected in parallel, then the equivalent
resistance is,

$$R_1 = \frac{3 \times 6}{3 + 6} = 2\,\Omega \tag{1.141}$$

Again, 4 and 2 $\Omega$ resistances are in series, then parallel with 12 $\Omega$ resistance. So,
the equivalent resistance is calculated as,

$$R_2 = \frac{(4 + 2) \times 12}{4 + 2 + 12} = 4\,\Omega \tag{1.142}$$

**Fig. 1.40** Circuit for
Example 1.18

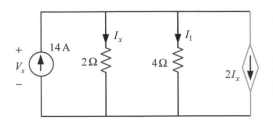

Finally, 4 and 2 Ω are in series and parallel with 12 Ω resistance. Then the total circuit resistance is calculated as,

$$R_t = 8 + \frac{(4+2) \times 12}{4+2+12} = 12\,\Omega \tag{1.143}$$

The value of the source current is calculated as,

$$I_s = \frac{48}{12} = 4\,\text{A} \tag{1.144}$$

**Example 1.18** A circuit with a current-controlled current source is shown in Fig. 1.40. Determine the branch currents and power absorbed by the 4 Ω resistance.

**Solution**
Applying KCL to the circuit in Fig. 1.40 yields,

$$14 = I_x + I_1 + 2I_x \tag{1.145}$$

Applying Ohm's law to the circuit in Fig. 1.39 yields,

$$I_x = \frac{V_s}{2} \tag{1.146}$$

$$I_1 = \frac{V_s}{4} \tag{1.147}$$

Substituting Eqs. (1.146) and (1.147) into Eq. (1.145) yields,

$$14 = 3\frac{V_s}{2} + \frac{V_s}{4} \tag{1.148}$$

$$14 = \frac{7V_s}{4} \tag{1.149}$$

$$V_s = \frac{14 \times 4}{7} = 8\,\text{V} \tag{1.150}$$

Substituting Eq. (1.150) into Eqs. (1.146) and (1.147) yields,

$$I_x = \frac{8}{2} = 4\,\text{A} \tag{1.151}$$

$$I_1 = \frac{8}{4} = 2\,\text{A} \tag{1.152}$$

The power absorbed by the 4 $\Omega$ resistance is calculated as,

$$P_{4\Omega} = 2^2 \times 4 = 16\,\text{W} \tag{1.153}$$

### Practice Problem 1.16
A series–parallel circuit is shown in Fig. 2.21. Calculate the total resistance and the current through 8 $\Omega$ resistance (Fig. 1.41).

### Practice Problem 1.17
A series–parallel circuit with two jumper wires is shown in Fig. 1.42. Determine the total resistance and the source current.

### Practice Problem 1.18
A circuit with a voltage-controlled current source is shown in Fig. 1.43. Calculate the current $I_1$ and the power absorbed by the 3 $\Omega$ resistance.

**Fig. 1.41** Circuit for
Practice Problem 1.16

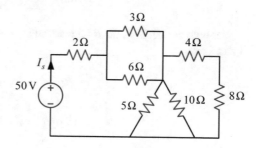

**Fig. 1.42** Circuit for
Practice Problem 1.17

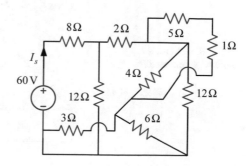

**Fig. 1.43** Circuit for
Practice Problem 1.18

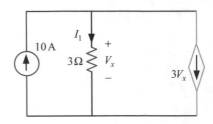

## 1.17  Delta-Wye Transformation

Resistances are sometimes either not connected in series or parallel. In this case, a different technique for reducing electrical circuits is required. This technique is known as delta-wye or wye-delta transformation. Consider three resistances that are connected in delta and wye as shown in Fig. 1.44.

For delta connection, the total resistances between terminals 1 and 2, 2 and 3, and 3 and 1 are,

$$R_{12(\Delta)} = \frac{R_b(R_a + R_c)}{R_a + R_b + R_c} \tag{1.154}$$

$$R_{23(\Delta)} = \frac{R_c(R_a + R_b)}{R_a + R_b + R_c} \tag{1.155}$$

$$R_{31(\Delta)} = \frac{R_a(R_b + R_c)}{R_a + R_b + R_c} \tag{1.156}$$

For wye connection, the total resistances between terminals 1 and 2, 2 and 3, and 3 and 1 are,

$$R_{12(Y)} = R_1 + R_2 \tag{1.157}$$

$$R_{23(Y)} = R_2 + R_3 \tag{1.158}$$

$$R_{31(Y)} = R_3 + R_1 \tag{1.159}$$

**Fig. 1.44** Circuit for
delta-wye conversion

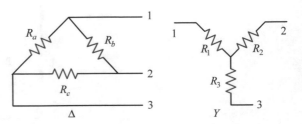

Electrically, the total resistances between terminals 1 and 2, 2 and 3, and 3 and 1 are the same for delta and wye connections. Then, the following relations can be written as,

$$R_{12(Y)} = R_{12(\Delta)} \tag{1.160}$$

$$R_{23(Y)} = R_{23(\Delta)} \tag{1.161}$$

$$R_{31(Y)} = R_{31(\Delta)} \tag{1.162}$$

Substituting Eqs. (1.154) and (1.157) into Eq. (1.160) yields,

$$R_1 + R_2 = \frac{R_b(R_a + R_c)}{R_a + R_b + R_c} \tag{1.163}$$

Substituting Eqs. (1.155) and (1.158) into Eq. (1.161) yields,

$$R_2 + R_3 = \frac{R_c(R_a + R_b)}{R_a + R_b + R_c} \tag{1.164}$$

Again, substituting Eqs. (1.156) and (1.159) into Eq. (1.162) yields,

$$R_3 + R_1 = \frac{R_a(R_b + R_c)}{R_a + R_b + R_c} \tag{1.165}$$

Adding Eqs. (1.163), (1.164) and (1.165) yields,

$$2(R_1 + R_2 + R_3) = \frac{2(R_a R_b + R_b R_c + R_c R_a)}{R_a + R_b + R_c} \tag{1.166}$$

$$R_1 + R_2 + R_3 = \frac{R_a R_b + R_b R_c + R_c R_a}{R_a + R_b + R_c} \tag{1.167}$$

Subtracting Eqs. (1.164), (1.165) and (1.163) from Eq. (1.167) yields,

$$R_1 = \frac{R_a R_b}{R_a + R_b + R_c} \tag{1.168}$$

$$R_2 = \frac{R_b R_c}{R_a + R_b + R_c} \tag{1.169}$$

$$R_3 = \frac{R_c R_a}{R_a + R_b + R_c} \tag{1.170}$$

## 1.18  Wye-Delta Transformation

Resistances are in wye connection required to transform into delta connection for circuit reduction. The formulae for this transformation can be derived from Eqs. (1.168), (1.169) and (1.170). Multiplying Eq. (1.168) by (1.169), Eq. (1.169) by (1.170) and Eq. (1.170) by (1.168) yields,

$$R_1 R_2 = \frac{R_a R_b^2 R_c}{(R_a + R_b + R_c)^2} \tag{1.171}$$

$$R_2 R_3 = \frac{R_a R_b R_c^2}{(R_a + R_b + R_c)^2} \tag{1.172}$$

$$R_3 R_1 = \frac{R_a^2 R_b R_c}{(R_a + R_b + R_c)^2} \tag{1.173}$$

Adding Eqs. (1.171), (1.172) and (1.173) yields,

$$R_1 R_2 + R_2 R_3 + R_3 R_1 = \frac{R_a R_b^2 R_c + R_a R_b R_c^2 + R_a^2 R_b R_c}{(R_a + R_b + R_c)^2} \tag{1.174}$$

$$R_1 R_2 + R_2 R_3 + R_3 R_1 = \frac{R_a R_b R_c (R_a + R_b + R_c)}{(R_a + R_b + R_c)^2} \tag{1.175}$$

$$R_1 R_2 + R_2 R_3 + R_3 R_1 = \frac{R_a R_b R_c}{R_a + R_b + R_c} \tag{1.176}$$

Dividing Eq. (1.176) by Eqs. (1.169), (1.170) and (1.168) yields,

$$R_a = \frac{R_1 R_2 + R_2 R_3 + R_3 R_1}{R_2} \tag{1.177}$$

$$R_b = \frac{R_1 R_2 + R_2 R_3 + R_3 R_1}{R_3} \tag{1.178}$$

$$R_c = \frac{R_1 R_2 + R_2 R_3 + R_3 R_1}{R_1} \tag{1.179}$$

*Example 1.19* A delta-wye circuit is shown in Fig. 1.45. Calculate the total circuit resistance, the source current and the power absorbed by the 4 Ω resistance.

**Solution**
Converting the upper part delta circuit to wye circuit as shown in Fig. 1.46. The values of the resistances are calculated as,

**Fig. 1.45**  Circuit for
Example 1.19

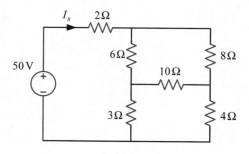

**Fig. 1.46**  Circuit converted
to wye

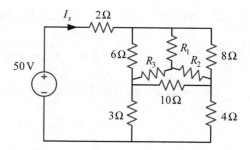

$$R_1 = \frac{6 \times 8}{6 + 10 + 8} = 2\,\Omega \tag{1.180}$$

$$R_2 = \frac{10 \times 8}{6 + 10 + 8} = 3.33\,\Omega \tag{1.181}$$

$$R_3 = \frac{10 \times 6}{6 + 10 + 8} = 2.5\,\Omega \tag{1.182}$$

Now, the circuit in Fig. 1.46 can be re-drawn as shown in Fig. 1.47. The $R_3$ and
$3\,\Omega$ are in series and $R_2$ and $4\,\Omega$ are in series, then parallel each other. In this case,
the equivalent resistance is calculated as,

$$R_4 = \frac{(R_2 + 4)(R_3 + 3)}{R_2 + 4 + R_3 + 3} = \frac{(3.33 + 4)(2.5 + 3)}{3.33 + 4 + 2.5 + 3} = 3.14\,\Omega \tag{1.183}$$

From Fig. 1.47, the total circuit resistance is determined as,

$$R_t = 2 + R_1 + R_4 = 2 + 2 + 3.14 = 7.14\,\Omega \tag{1.184}$$

The value of the source current is calculated as,

$$I_s = \frac{50}{7.14} = 7\,\text{A} \tag{1.185}$$

**Fig. 1.47** Final circuit

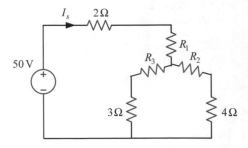

The current through 4 Ω resistance is calculated as,

$$I_{4\Omega} = 7 \times \frac{(3 + 2.5)}{5.5 + 3.33 + 4} = 3\,\text{A} \tag{1.186}$$

The power absorbed by 4 Ω resistance is calculated as,

$$P_{4\Omega} = I_{4\Omega}^2 \times 4 = 3^2 \times 4 = 36\,\text{W} \tag{1.187}$$

***Example 1.20*** Figure 1.48 shows a delta-wye circuit. Determine the total circuit resistance, and the source current.

**Solution**
Converting the wye connection to a delta connection as shown in Fig. 1.49. The values of the resistances are calculated as,

$$R_1 = \frac{7 \times 10 + 10 \times 3 + 3 \times 7}{3} = 40.33\,\Omega \tag{1.188}$$

$$R_2 = \frac{7 \times 10 + 10 \times 3 + 3 \times 7}{7} = 17.29\,\Omega \tag{1.189}$$

$$R_3 = \frac{7 \times 10 + 10 \times 3 + 3 \times 7}{10} = 12.1\,\Omega \tag{1.190}$$

**Fig. 1.48** Circuit for Example 1.20

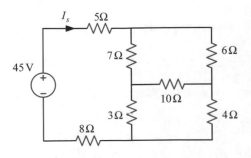

**Fig. 1.49** Circuit from wye to delta

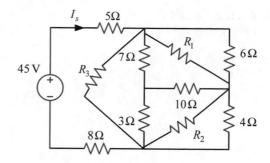

From Fig. 1.50, it is seen that $R_1$ and 6 $\Omega$ are in parallel, and $R_2$ and 4 $\Omega$ are in parallel, then they are in series. The value of the resistance is calculated as,

$$R_4 = \frac{R_1 \times 6}{R_1 + 6} + \frac{R_2 \times 4}{R_2 + 4} = \frac{40.33 \times 6}{40.33 + 6} + \frac{17.29 \times 4}{17.29 + 4} = 8.47 \,\Omega \qquad (1.191)$$

The $R_3$ and 8.47 $\Omega$ resistances are in parallel, then series with 5 and 8 $\Omega$ resistances as shown in Fig. 1.51. Finally, the total circuit resistance is calculated as,

$$R_t = \frac{12.1 \times 8.47}{12.1 + 8.47} + 5 + 8 = 17.98 \,\Omega \qquad (1.192)$$

The value of the source current can be determined as,

**Fig. 1.50** Circuit after wye to delta conversion

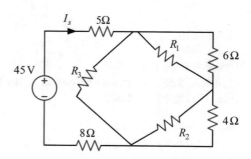

**Fig. 1.51** Circuit for the final calculation

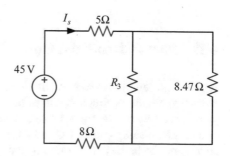

**Fig. 1.52** Circuit for
practice Problem 1.19

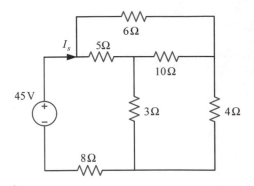

**Fig. 1.53** Circuit for
Practice Problem 1.20

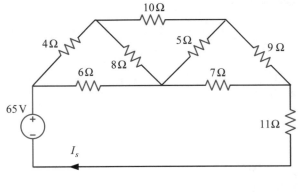

$$I_s = \frac{45}{17.98} = 2.5\,\text{A} \tag{1.193}$$

**Practice Problem 1.19**
Figure 2.32 shows a delta-wye circuit. Calculate the total circuit resistance, and the
source current (Fig. 1.52).

**Practice Problem 1.20**
Figure 1.53 shows a delta-wye circuit. Determine the total circuit resistance, and the
source current.

## 1.19    Short Circuit and Open Circuit

A circuit is identified as a short circuit when a living object is becoming a part of the
energized circuit. A circuit is said to be a short circuit when current flows through a
path with a very low (zero) resistance. The voltage is zero in a short circuit. The short
circuit is happening both in DC and AC circuits. When two terminals of a dry cell or
an automobile battery are shorted by a piece of wire, then the battery will discharge

**Fig. 1.54**  A resistance
across a short wire

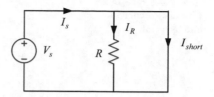

through the wire, which in turn heat the wire. The principle of arc welding is based
on the application of heating due to a short circuit. Figure 1.54 shows a circuit to
explain the short circuit phenomena. Applying the current divider rule to find the
currents in both branches. These currents are,

$$I_R = I_s \times \frac{0}{R+0} = 0 \tag{1.194}$$

$$I_{short} = I_s \times \frac{R}{R+0} = I_s \tag{1.195}$$

The voltage in the short circuit is calculated as,

$$V_{short} = I_{short} \times R_{short} = I_s \times 0 = 0 \tag{1.196}$$

A circuit is said to be an open circuit when it contains infinite resistance.
Figure 1.55 illustrates the concept of an open circuit. From Fig. 2.34, the expression
of the current can be written as,

$$I_s = \frac{V_s}{R} = \frac{V_s}{\infty} = 0 \tag{1.197}$$

**Example 1.21**  A short and open circuit is shown in Fig. 1.56. Determine the total
circuit resistance, and the source current when terminals $a$ and $b$ are short and open
circuits.

**Fig. 1.55**  A circuit with
open terminals

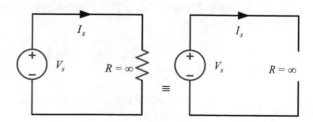

**Fig. 1.56** A circuit for
Example 1.21

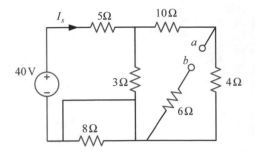

**Solution**

Consider the terminals $a$ and $b$ are short-circuited. Then, the resistances 6 and 4 $\Omega$ are in parallel, then series with 10 $\Omega$ resistance. The value of the resistance is calculated as,

$$R_1 = \frac{4 \times 6}{4 + 6} + 10 = 12.4\,\Omega \tag{1.198}$$

Again, 12.4 and 3 $\Omega$ are in parallel. In this case, the value of the resistance is calculated as,

$$R_2 = \frac{12.4 \times 3}{12.4 + 3} = 2.42\,\Omega \tag{1.199}$$

The total circuit resistance is calculated as,

$$R_{ts} = 5 + 2.42 = 7.42\,\Omega \tag{1.200}$$

The value of the source current is calculated as,

$$I_{ss} = \frac{40}{7.42} = 5.39\,\text{A} \tag{1.201}$$

Again, consider the terminals $a$ and $b$ are open circuited. The resistances 4 and 10 $\Omega$ are in series, then parallel with 3 $\Omega$ resistance. The values of the resistance are calculated as,

$$R_3 = 4 + 10 = 14\,\Omega \tag{1.202}$$

$$R_4 = \frac{14 \times 3}{14 + 3} = 2.47\,\Omega \tag{1.203}$$

The total circuit resistance is calculated as,

$$R_{to} = 5 + 2.47 = 7.47\,\Omega \tag{1.204}$$

**Fig. 1.57** A circuit for
Practice Problem 1.21

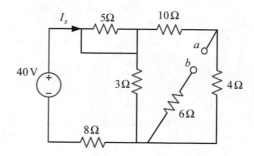

The value of the source current is calculated as,

$$I_{so} = \frac{40}{7.47} = 5.35 \, \text{A} \tag{1.205}$$

**Practice Problem 1.21**

Figure 1.57 shows a short and open circuit. Find the total circuit resistance and the source current.

## 1.20 Source Configuration

The voltage and current sources are usually connected either in series or parallel configurations. There are six cells in an automobile battery. Each cell contains two volts and they are connected in a series connection to generate twelve volts. Two voltage sources are connected in series as shown in Fig. 1.58.

Consider the voltage source $V_{s1}$ is greater than the voltage source $V_{s2}$. Then the following equations from Fig. 1.58 can be written as,

$$V_a = V_{s1} + V_{s2} \tag{1.206}$$

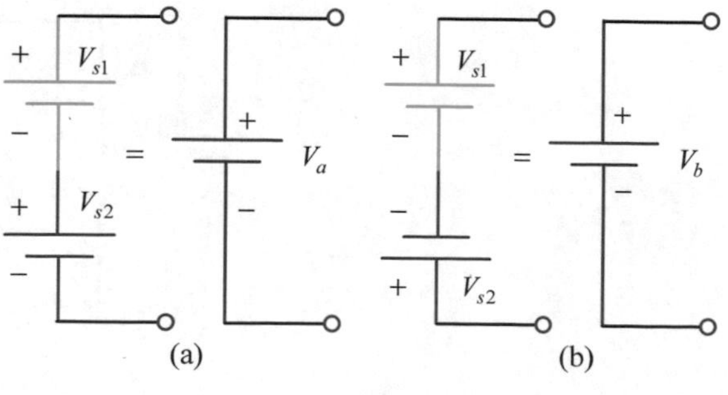

**Fig. 1.58** A circuit with
voltage sources

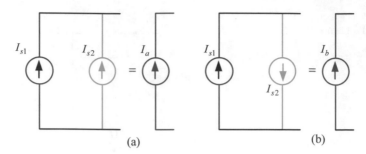

**Fig. 1.59** A circuit with current sources

$$V_b = V_{s1} - V_{s2} \qquad (1.207)$$

The current sources are not recommended for series connection, but they are usually connected in a parallel in the circuit as shown in Fig. 1.59.

Consider the current source $I_{s1}$ is greater than the current source $I_{s2}$. From Fig. 1.59, the following equations can be written as,

$$I_a = I_{s1} + I_{s2} \qquad (1.208)$$

$$I_b = I_{s1} - I_{s2} \qquad (1.209)$$

**Example 1.22** A series–parallel circuit with two voltage sources is shown in Fig. 1.60. Calculate the total circuit resistance, source current and the power absorbed by 10 Ω resistance.

**Solution**
Two voltage sources are connected in series but in opposite direction. Therefore, the value of the resultant voltage is calculated as,

**Fig. 1.60** A circuit for
Example 1.22

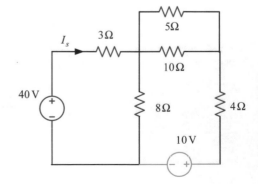

$$V_s = 40 - 10 = 30 \, \text{V} \tag{1.210}$$

The resistances 5 and 10 $\Omega$ are in parallel, then in series with 4 $\Omega$. The value of the resistance is calculated as,

$$R_1 = \frac{5 \times 10}{5 + 10} + 4 = 7.33 \, \Omega \tag{1.211}$$

The 7.33 $\Omega$ resistance is in parallel with 8 $\Omega$ resistance, then series with 3 $\Omega$ resistance. The total circuit resistance is calculated as,

$$R_t = \frac{7.33 \times 8}{7.33 + 8} + 3 = 6.83 \, \Omega \tag{1.212}$$

The value of the source current is calculated as,

$$I_s = \frac{30}{6.83} = 4.39 \, \Omega \tag{1.213}$$

The current in the 4 $\Omega$ resistance is calculated as,

$$I_{4\Omega} = 4.39 \times \frac{8}{8 + 4 + 3.33} = 2.29 \, \text{A} \tag{1.214}$$

The current through 10 $\Omega$ resistance is calculated as,

$$I_{10\Omega} = 2.29 \times \frac{5}{5 + 10} = 0.76 \, \text{A} \tag{1.215}$$

Power absorbed by 10 $\Omega$ resistance is calculated as,

$$P_{10\Omega} = 0.76^2 \times 10 = 5.78 \, \text{W} \tag{1.216}$$

**Example 1.23** A parallel circuit with two current sources is shown in Fig. 1.61. Find the current in the 8 $\Omega$ resistance.

**Fig. 1.61** A circuit for Example 1.23

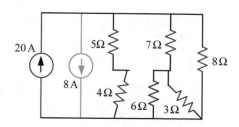

**Solution**
The value of the equivalent current source is calculated as,

$$I_s = 20 - 8 = 12\,\text{A} \tag{1.217}$$

The 6 and 3 $\Omega$ resistances are in parallel, then series with 7 $\Omega$ resistance. Here, the equivalent resistance is calculated as,

$$R_1 = \frac{3 \times 6}{3 + 6} + 7 = 9\,\Omega \tag{1.218}$$

Then 9 $\Omega$ is in parallel with $5 + 4 = 9\,\Omega$ resistance, and the value of the equivalent resistance is calculated as,

$$R_2 = \frac{9 \times 9}{9 + 9} = 4.5\,\Omega \tag{1.219}$$

The current through 8 $\Omega$ resistance is calculated as,

$$I_{8\Omega} = 12 \times \frac{4.5}{4.5 + 8} = 4.32\,\text{A} \tag{1.220}$$

**Practice Problem 1.22**
Figure 1.62 shows a series–parallel circuit with two voltage sources. Determine the current in the 6 $\Omega$ resistance.

**Practice Problem 1.23**
Figure 1.63 shows a parallel circuit with two current sources. Determine the current in the 8 $\Omega$ resistance.

**Exercise Problems**

1.1   A charge with an expression of $q = e^{-2t} + 6t^2 + 3t + 2$ C is found in a circuit terminal. Determine the general expression of current and its value at $t = 0.01$ s.

**Fig. 1.62** A circuit for
Practice Problem 1.22

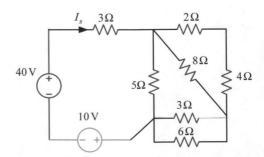

**Fig. 1.63** A circuit for
Practice Problem 1.23

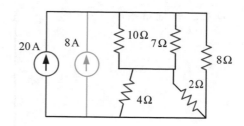

**Fig. P1.1** Waveform for
problem 1.4

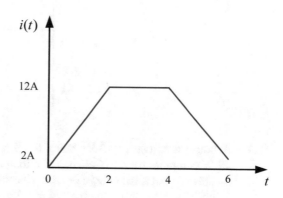

1.2   A current of 6 A passes through a wire, where the expression of charge is found
      to be $q = e^{2t} + 3t$ C. Calculate the value of the time.

1.3   A charge of 8 C moves through a wire for 2 s. Determine the value of the
      current.

1.4   A current as shown in Fig. P1.1 passes through a wire. Find the expressions
      of current for different times and the value of the charge.

1.5   Fig. P1.2 shows a current waveform that passes through a wire. Calculate the
      expression of current and the value of the associated charge.

1.6   The resistivity of a 200 m long and $1.04 \times 10^{-6}$ m² copper wire is $1.72 \times 10^{-8}$
      Ω-m. Calculate the value of the resistance.

1.7   The resistance of a copper wire is found to be 10 Ω. If the area and the resistivity
      of the wire are $0.95 \times 10^{-7}$ m² and $1.72 \times 10^{-8}$ Ω-m, respectively, then find
      the length of the wire.

1.8   The resistivity of a 200 m long aluminum wire is $2.8 \times 10^{-8}$ Ω-m. Determine
      the diameter of the wire, if the resistance of the wire is 6Ω.

1.9   A work with an expression of $w = 5q^3 - 3q^2 + q$ J is required to move a
      charge from one point to another point of a conductor. Calculate the voltage
      when $q = 0.01$ C.

1.10  The work done for a time $t$ s is given by $w = 2t^2 + 10t$ J. Find the value of
      the time for the corresponding power of 26 W.

1.11  A power $p = 2e^{-2t} + 5$ W is used for a time period from 0.01 s to 0.03 s to
      complete a task. Calculate the corresponding energy.

**Fig. P1.2**  Waveform for
problem 1.5

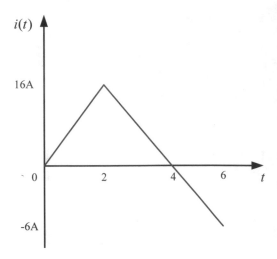

1.12  A house uses twelve 40 W bulbs for 3 h, a 730 W iron for 1.5 h and a 40 W computer for 5 h. Calculate the total energy used in the house and the associated cost if the energy price is 10 cents per kWh.

1.13  A 750 W electric iron is connected to a 220 V voltage source. Determine the current draws by an electric iron.

1.14  A 15 Ω resistance is connected to a 240 V voltage source. Calculate the power absorbed by the resistance.

1.15  A circuit with different branches along with currents is shown in Fig. P1.3. Determine the values of unknown currents.

1.16  Fig. P1.4 shows a circuit with some unknown currents. Determine the values of unknown currents.

1.17  Some known and unknown voltages of a circuit are shown in Fig. P1.5. Use KVL to determine unknown voltages.

1.18  Fig. P1.6 shows a circuit with known and unknown voltages. Use KVL to find the voltages $V_1$ and $V_2$.

1.19  A voltage-controlled voltage source is shown in Fig. P1.6. Use KVL to determine the voltage $V_x$ and the voltage between the two points $a$ and $b$ (Fig. P1.7).

**Fig. P1.3**  Circuit for
problem 1.15

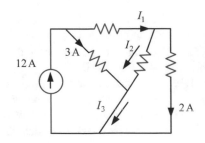

**Fig. P1.4**  Circuit for
problem 1.16

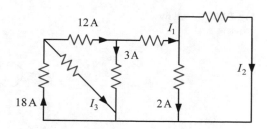

**Fig. P1.5**  Circuit for
problem 1.17

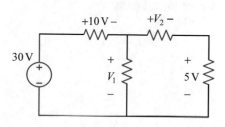

**Fig. P1.6**  Circuit for
problem 1.18

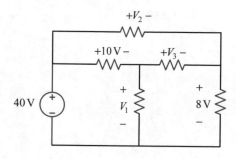

**Fig. P1.7**  Circuit for
problem 1.19

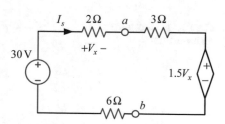

1.20  A current-controlled voltage source is shown in Fig. P1.8. Determine the
currents $I_s$ and $I_x$ using KVL.

1.21  A voltage-controlled current source is shown in Fig. P1.9. Calculate the
voltage drop by the 4 $\Omega$ resistance.

1.22  Fig. P1.10 shows a current-controlled current source. Use KCL to determine
the power absorbed by the 8 $\Omega$ resistance.

1.23  A series–parallel circuit with a voltage source is shown in Fig. P1.11.
Calculate the total resistance and the source current.

**Fig. P1.8**  Circuit for
problem 1.20

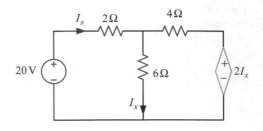

**Fig. P1.9**  Circuit for
problem 1.21

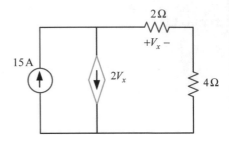

**Fig. P1.10**  Circuit for
problem 1.22

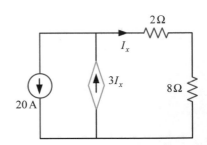

**Fig. P1.11**  Circuit for
problem 1.23

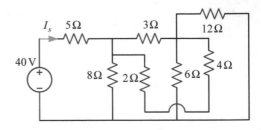

1.24    Fig. P1.12 shows a series–parallel circuit with a voltage source. Calculate the
        total circuit resistance and the source current.

1.25    A series–parallel circuit with a voltage source is shown in Fig. P1.13.
        Determine the total circuit resistance and the source current.

1.26    A series–parallel circuit is shown in Fig. P1.14. Calculate the total circuit
        resistance and the power absorbed by the 4 Ω resistance.

**Fig. P1.12** Circuit for
problem 1.24

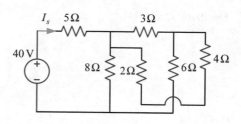

**Fig. P1.13** Circuit for
problem 1.25

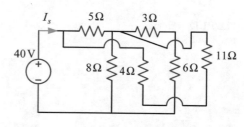

**Fig. P1.14** Circuit for
problem 1.26

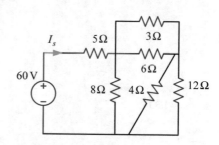

1.27   Find the current through 4 Ω resistance of a series–parallel circuit as shown
in Fig. P1.15.

1.28   A series–parallel circuit is shown in Fig. P1.16. Calculate the source current
and the voltage drop by the 4 Ω resistance.

1.29   Fig. P1.17 shows a series–parallel circuit. Determine the source current and
the voltage drop by the 8 Ω resistance.

1.30   Fig. P1.18 shows a series–parallel circuit. Find the total circuit resistance and
the source current.

**Fig. P1.15** Circuit for
problem 1.27

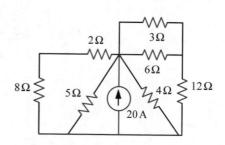

**Fig. P1.16**  Circuit for
problem 1.28

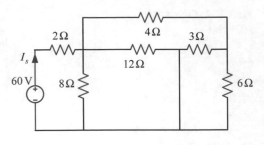

**Fig. P1.17**  Circuit for
problem 1.29

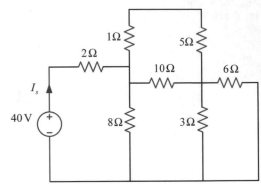

**Fig. P1.18**  Circuit for
problem 1.30

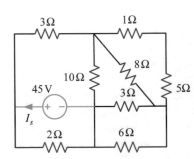

**Fig. P1.19**  Circuit for
problem 1.31

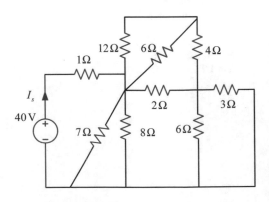

1.31  A series–parallel circuit is shown in Fig. P1.19. Calculate the total circuit resistance and the source current.

1.32  Fig. P1.20 shows a series–parallel circuit. Calculate the total circuit resistance and the source current.

1.33  A series–parallel electrical circuit is shown in Fig. P1.21. Find the total circuit resistance and the source current.

1.34  Fig. P1.22 shows a series–parallel electrical circuit. Calculate the total circuit resistance and the source current.

**Fig. P1.20** Circuit for problem 1.32

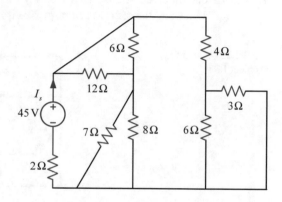

**Fig. P1.21** Circuit for problem 1.33

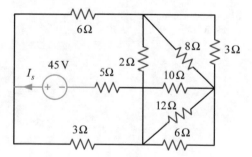

**Fig. P1.22** Circuit for problem 1.34

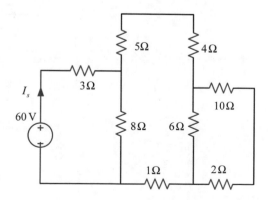

1.35    A series–parallel circuit is shown in Fig. P1.23. Determine the total circuit
        resistance and the source current.

1.36    A series–parallel circuit is shown in Fig. P1.24. Calculate the total circuit
        resistance and the source current.

1.37    An electrical circuit is shown in Fig. P1.25. Determine the total circuit
        resistance and the power absorbed by the 2 Ω resistance.

1.38    Fig. P1.26 shows an electrical circuit. Calculate the total circuit resistance
        and the current through the 6 Ω resistance.

1.39    An electrical circuit is shown in Fig. P1.27. Calculate the total circuit
        resistance and the source current in the circuit.

**Fig. P1.23**  Circuit for
problem 1.35

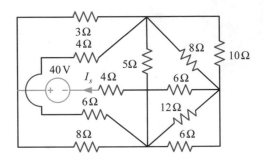

**Fig. P1.24**  Circuit for
problem 1.36

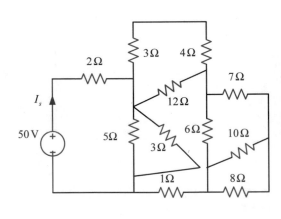

**Fig. P1.25**  Circuit for
problem 1.37

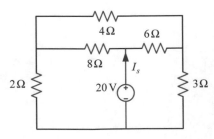

**Fig. P1.26** Circuit for
problem 1.38

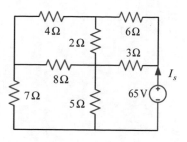

**Fig. P1.27** Circuit for
problem 1.39

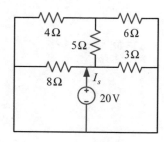

**Fig. P1.28** Circuit for
problem 1.40

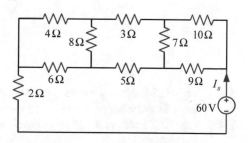

1.40    A delta-wye electrical circuit is shown in Fig. P1.28. Determine the total
        circuit resistance, source current and the voltage drop by the 3 Ω resistance.
1.41    An electrical circuit is shown in Fig. P1.29. Use delta-wye conversion to
        determine the total circuit resistance and the source current in the circuit.
1.41    A series-parallel electrical circuit is shown in Fig. P1.30. Calculate the power
        absorbed by the 4 Ω resistance.

**Fig. P1.29** Circuit for
problem 1.41

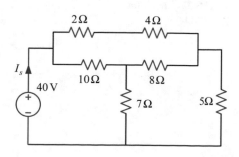

**Fig. P1.30**  Circuit for
problem 1.41

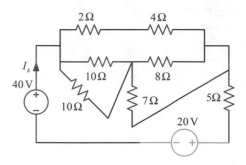

**Fig. P1.31**  Circuit for
problem 1.42

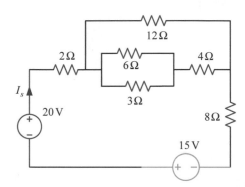

1.42   Fig. P1.31 shows a series–parallel electrical circuit. Calculate the current in
the 4 Ω resistance.

1.43   Fig. P1.32 shows a series–parallel electrical circuit. Calculate the value of
the voltage $V_x$.

1.44   A series–parallel electrical circuit with two current sources is shown in
Fig. P1.33. Determine the current in the 5 Ω resistance.

1.45   A series–parallel electrical circuit is shown in Fig. P1.34. Calculate the
voltage across the 4 Ω resistance.

1.46   Figure P1.35 shows a series–parallel electrical circuit. Determine the power
absorbed by the 10 Ω resistance.

**Fig. P1.32**  Circuit for
problem 1.43

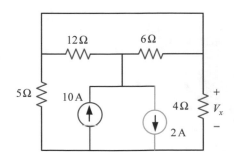

**Fig. P1.33** Circuit for
problem 1.44

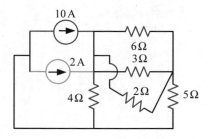

**Fig. P1.34** Circuit for
problem 1.45

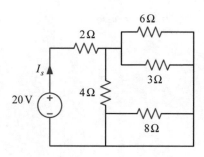

**Fig. P1.35** Circuit for
problem 1.46

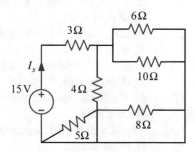

1.47   Figure P1.36 shows a series–parallel electrical circuit. Determine the value
        of the voltage, $V_x$.

1.48   Calculate the value of the voltage, $V_0$ of the circuit as shown in Fig. P1.37.

**Fig. P1.36** Circuit for
problem 1.47

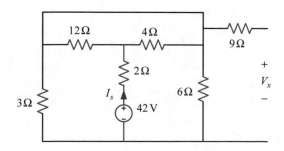

**Fig. P1.37** Circuit for problem 1.48

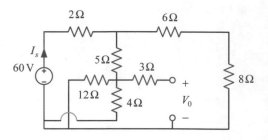

**Fig. P1.38** Circuit for problem 1.49

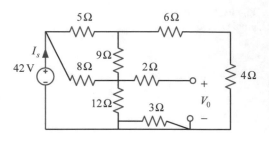

1.49    An electrical circuit is shown in Fig. P1.38. Calculate the value of the voltage, $V_0$ of the circuit.

1.50    An electrical circuit is shown in Fig. P1.39. Determine the value of the voltage, $V_0$ of the circuit.

1.51    Figure P1.40 shows an electrical circuit. Calculate the value of the voltage, $V_0$ of the circuit.

1.52    An electrical circuit is shown in Fig. P1.41. Find the value of the voltage, $V_0$ of the circuit.

1.53    An electrical circuit is shown in Fig. P1.42. Determine the value of the voltage, $V_0$ of the circuit.

1.54    An electrical circuit is shown in Fig. P1.43. Calculate the value of the voltage, $V_0$ of the circuit.

1.55    Figure P1.44 shows an electrical circuit. Determine the value of the voltage, $V_x$ of the circuit.

**Fig. P1.39** Circuit for problem 1.50

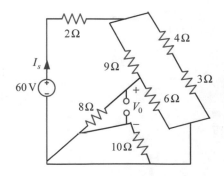

**Fig. P1.40** Circuit for
problem 1.51

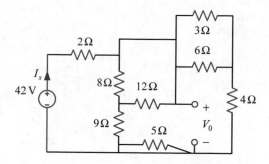

**Fig. P1.41** Circuit for
problem 1.52

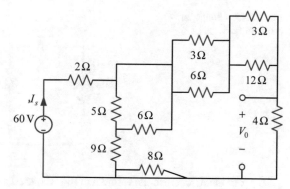

**Fig. P1.42** Circuit for
problem 1.53

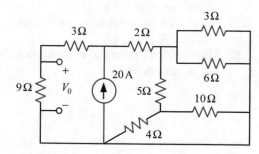

**Fig. P1.43** Circuit for
problem 1.54

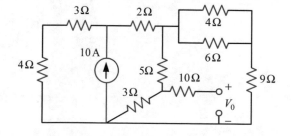

**Fig. P1.44**  Circuit for
problem 1.55

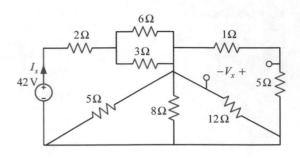

**Fig. P1.45**  Circuit for
problem 1.56

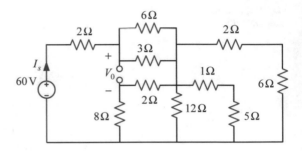

1.56    An electrical circuit is shown in Fig. P1.45. Calculate the value of the voltage,
        $V_0$ of the circuit.
1.57    Figure P1.46 shows an electrical circuit. Find the value of the voltage, $V_0$ of
        the circuit.
1.58    Figure P1.47 shows an electrical circuit. Calculate the value of the voltage,
        $V_0$ of the circuit.

**Fig. P1.46**  Circuit for
problem 1.57

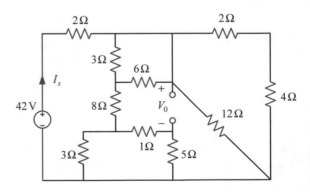

**Fig. P1.47** Circuit for problem 1.58

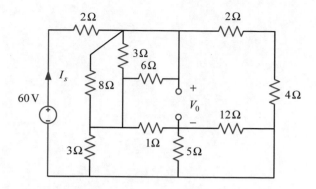

# References

1. C.K. Alexander, M.N.O. Sadiku, *Fundamentals of Electric Circuits*, 6th edn. (McGraw-Hill Higher Education, 2016)
2. J. David Irwin, R. Mark Nelms *Basic Engineering Circuit Analysis*, 11th edn. (Wiley, USA, 2015)
3. J.W. Nilsson, S.A. Riedel, *Electric Circuits*, 10th edn. (Prentice-Hall International Edition, 2015)
4. R.L. Boylestad, *Introductory Circuit Analysis*, 13th edn. (Pearson, 2016)
5. M.A. Salam, *Basic Electrical Circuits*, 2nd edn. (Shroff Publishers & Distributors Pvt. Ltd, India, 2007)
6. B. Anderson, R. Anderson, *Fundamentals of Semiconductor Devices*, 2nd edn. (McGraw-Hill Higher Education, 2017)

# Chapter 2
# Analysis of Electrical Power

## 2.1 Introduction

In Chap. 1, the analysis has been limited to the electrical-circuit networks with time-invariant sources, also known as DC sources. At this point, the focus of the circuit analysis will be shifted towards time-varying sources. In the electrical domain, the magnitude of a time-varying source varies between two preset levels with time. In this domain, a sinusoidal time-varying source is generally known as AC (derived from the name, Alternating Current) source. Circuits, driven by AC sources (current or voltage), are known as AC circuits. Generation, transmission and distribution of AC sources are more efficient than time-invariant sources. AC signals in any electrical circuit are also easy to analyze with diverse mathematical tools. The development of AC sources along with the formulation of associated mathematical theories has expanded the electrical power industries at an enormous rate in the whole world. AC generator, transformer, and other related high voltage devices are required to generate, transmit and distribute alternating voltage and current. These types of equipment are rated by Mega Watt (MW) and Mega Voltage-Ampere (MVA). This chapter will discuss instantaneous power, average power, complex power, power factor, maximum power transfer theorem, and power factor correction.

## 2.2 Waveforms

The sinusoidal source generates voltage or current that varies with time. A sinusoidal signal has the form of either sine or cosine function [1, 2], as shown in Fig. 2.1 (in the form of a sine function). Other than sinusoidal signals, AC circuits deal with a few more time-varying signal waveforms. These are triangular, saw-tooth, and pulse waveforms. The waveform whose rate of rising and the rate of fall is equal and constant is known as a triangular waveform. The waveform whose magnitude varies

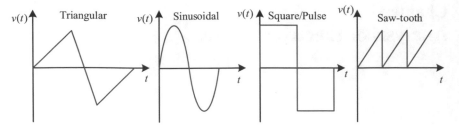

**Fig. 2.1** Different types of waveforms

between two constant positive and negative magnitudes is known as pulse or square waveform.

The waveform whose rate of rising is different from the rate of fall is known as the saw-tooth waveform (looks like saw-tooth). All these waveforms are shown in Fig. 2.1. The value of the waveform at any instant of time is known as the instantaneous value. One complete set of positive and negative values of an alternating waveform is known as a cycle. The number of cycles that occur in one second is known as frequency. It is represented by the letter $f$ and its unit is cycles per second or hertz (Hz) in honour of the German physicist Heinrich Hertz (1857–1894). The total time required to complete one cycle of a waveform is known as the time period, which is denoted by the capital letter $T$. The period is inversely proportional to the frequency and it is written as,

$$T = \frac{1}{f}. \tag{2.1}$$

From Fig. 2.2, it is seen that an angular distance ($\omega t$) of $2\pi$ is traversed by the sinusoidal signal over a time period of $T$. In other words, the angular distance at $t = T$ is,

$$\omega T = 2\pi \tag{2.2}$$

**Fig. 2.2** Sinusoidal waveform

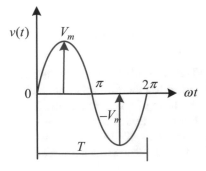

**Fig. 2.3** Two sinusoidal
voltage waveforms

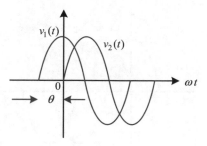

Substituting Eq. (2.1) into Eq. (2.2) yields,

$$\omega \frac{1}{f} = 2\pi \tag{2.3}$$

$$\omega = 2\pi f \tag{2.4}$$

In Eq. (2.4), $\omega$ is known as the angular frequency.

The highest magnitude of any waveform is known as the peak value or the maximum value. The maximum value of voltage and current are represented by $V_m$ for a voltage waveform and by $I_m$ for a current waveform. The sum of the positive and the negative peaks of a voltage waveform is known as the peak-to-peak value. From Fig. 2.2, the peak-to-peak value of the voltage waveform is written as,

$$V_{pp} = |V_m| + |-V_m| = 2V_m \tag{2.5}$$

The phase and phase differences between the two waveforms are very important to identify the terms lag and lead.

The phase (also known as phase-angle) is the initial angle of a sinusoidal function at its origin or a reference point. The phase difference is the difference in phase angles between the two waveforms as shown in Fig. 2.3.

From Fig. 2.3, the expressions of the two voltage waveforms can be written as,

$$v_1(t) = V_m \sin(\omega t + \theta) \tag{2.6}$$

$$v_2(t) = V_m \sin(\omega t + 0°) \tag{2.7}$$

The phase difference between the two waveforms can be written as,

$$\theta_{pd} = (\omega t + \theta) - (\omega t + 0°) = \theta \tag{2.8}$$

In an inductor, the current through the inductor lags the voltage across it by 90°, while in a capacitor the current leads the voltage by 90°. The lagging and leading phenomenon can be best understood by Figs. 2.4, respectively. In Fig. 2.4a, the

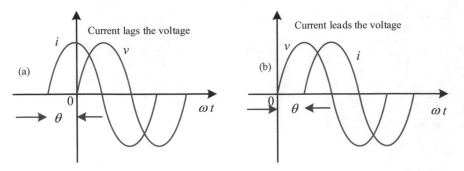

**Fig. 2.4** Lags and leads concept

current waveform $i$, lags the voltage waveform $v$, while in Fig. 2.4b, the current leads the voltage.

***Example 2.1*** The expression of an alternating voltage waveform is given by $v(t) = 50 \sin(314t - 35°)$ V. Determine the maximum value of the voltage, frequency, and time period.

**Solution:**
The maximum value of the voltage is,

$$V_m = 50\,\text{V} \qquad (2.9)$$

The frequency is calculated as,

$$\omega = 314 \qquad (2.10)$$

$$f = \frac{314}{2\pi} = 50\,\text{Hz} \qquad (2.11)$$

**Practice Problem 2.1**
The expression of an alternating current waveform is given by $i(t) = 15 \sin 377t$ A. Find the maximum value of the current, frequency, time period, and instantaneous current $t = 0.01$ s.

## 2.3   Root Mean Square Value

The root means square is an important term that is used in an AC circuit for analysis and it is abbreviated as rms. It is often known as effective value or virtual value. The rms value of an alternating current can be defined as the steady-state (DC) current

**Fig. 2.5** DC and AC sources
with resistance

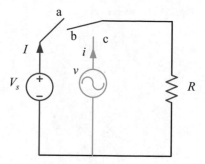

which, when flows through a given resistance for a given time-length produces the
same heat energy that is produced by an alternating current when flows through the
same resistance for the same length of time. According to the definition of rms value,
the following equation can be written as [3, 4],

$$P_{dc} = P_{ac} \tag{2.12}$$

A shown in the circuit in Fig. 2.5, when terminal 'a' connects to terminal 'b', the
heat energy produced due to the current $I$ is,

$$P_{dc} = I^2 R \tag{2.13}$$

Again, for the same circuit, when terminal 'b' connects the terminal 'c', the heat
energy produced due to the current $i$ is,

$$P_{ac} = \frac{1}{T} \int_0^T i^2(t) R dt \tag{2.14}$$

Substituting Eqs. (2.13) and (2.14) into Eq. (2.12) yields,

$$I^2 R = \frac{1}{T} \int_0^T i^2(t) R dt \tag{2.15}$$

$$I_{rms} = \sqrt{\frac{1}{T} \int_0^T i^2(t) dt} \tag{2.16}$$

Consider a sinusoidal current waveform as shown in Fig. 2.6. This sinusoidal
current is expressed as,

$$i = I_m \sin \omega t \tag{2.17}$$

**Fig. 2.6**  AC waveform

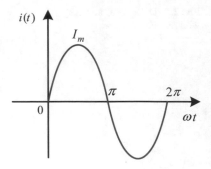

Substituting Eq. (2.17) into Eq. (2.16) yields,

$$I_{rms} = \sqrt{\frac{1}{2\pi} \int_0^{2\pi} (I_m \sin \omega t)^2 d(\omega t)} \tag{2.18}$$

$$I_{rms} = \sqrt{\frac{I_m^2}{4\pi} \int_0^{2\pi} 2 \sin^2 \omega t \, d(\omega t)} \tag{2.19}$$

$$I_{rms} = \frac{I_m}{2} \sqrt{\frac{1}{\pi} \int_0^{2\pi} (1 - \cos 2\omega t) d(\omega t)} \tag{2.20}$$

$$I_{rms} = \frac{I_m}{2} \sqrt{\frac{1}{\pi} [\omega t]_0^{2\pi} - \frac{1}{2\pi} [\sin 2\omega t]_0^{2\pi}} \tag{2.21}$$

$$I_{rms} = \frac{I_m}{2} \sqrt{\frac{1}{\pi} [2\pi - 0] - 0} \tag{2.22}$$

$$I_{rms} = \frac{I_m}{2} \times \sqrt{2} \tag{2.23}$$

$$I_{rms} = \frac{I_m}{\sqrt{2}} = 0.707 I_m \tag{2.24}$$

Equation (2.24) provides the expression for the rms current. Following a similar approach, the expression for the rms voltage can be derived as,

$$V_{rms} = \frac{V_m}{\sqrt{2}} = 0.707 V_m \tag{2.25}$$

In this case, the power absorbed by any resistance $R$ can be calculated as,

**Fig. 2.7** Waveform for
Example 2.2

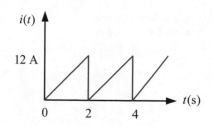

$$P = I_{rms}^2 R = \frac{V_{rms}^2}{R} \tag{2.26}$$

***Example 2.2*** A sawtooth current waveform is shown in Fig. 2.7. Calculate the rms value of this current, and.

when it flows through a 6 Ω resistor, calculate the power absorbed by the resistor.

**Solution:**

In the sawtooth waveform, the time-period is 2 s. The value of the slope is,

$$m = \frac{12}{2} = 6 \tag{2.27}$$

$$i(t) = 6t \text{ A} \quad 0 < t < 2 \tag{2.28}$$

The rms value of the current waveform is calculated as,

$$I_{rms}^2 = \frac{1}{2} \int_0^2 (6t)^2 dt \tag{2.29}$$

$$I_{rms}^2 = \frac{18}{3} \left[ t^3 \right]_0^2 = 6 \times 8 = 48 \tag{2.30}$$

$$I_{rms} = 6.92 \text{ A} \tag{2.31}$$

The power absorbed by the 6 Ω resistor is determined as,

$$P = 6.92^2 \times 6 = 288 \text{ W} \tag{2.32}$$

**Practice Problem 2.2**

Figure 2.8 shows a square-wave current waveform. Determine the rms value of the current, and the power absorbed by a 2 Ω resistor when the current flows through this resistor.

**Fig. 2.8** Waveform for
practice problem 2.2

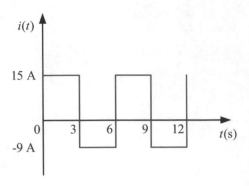

## 2.4   Average Value

The average of all the values of an alternating waveform in one cycle is known as the average value. The average value for a non-periodic waveform is expressed as,

$$\text{Average value} = \frac{\text{Area under one cycle}}{\text{base length}} \tag{2.33}$$

In general, the average value for a non-periodic waveform is written as,

$$I_{av} = \frac{1}{2\pi} \int_0^{2\pi} i\, d(\omega t) \tag{2.34}$$

For a symmetrical or periodic waveform, the positive area cancels the negative area. Thus, the average value is zero. In this case, the average value is calculated using the positive half of the waveform. The average value for a periodic waveform is expressed as,

$$\text{Average value} = \frac{\text{Area under half cycle}}{\text{base length}} \tag{2.35}$$

In general, the average value for a periodic waveform is written as,

$$I_{av} = \frac{1}{\pi} \int_0^{\pi} i\, d(\omega t) \tag{2.36}$$

The average value of the sinusoidal current waveform (periodic) is calculated as,

$$I_{av} = \frac{1}{\pi} \int_0^{\pi} I_m \sin \omega t\, d(\omega t) \tag{2.37}$$

**Fig. 2.9**  Waveform for
Example 2.3

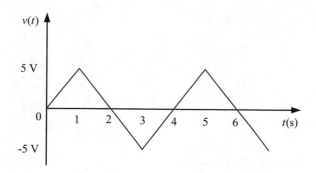

$$I_{av} = \frac{I_m}{\pi}[-\cos \omega t]_0^\pi = -\frac{I_m}{\pi}[\cos \pi - 1] \tag{2.38}$$

$$I_{av} = \frac{2I_m}{\pi} = 0.637 I_m \tag{2.39}$$

**Example 2.3**  A triangular current waveform is shown in Fig. 2.9. Find the average
value of this current waveform.

**Solution:**
The time-period of the waveform is 4 s. For different times, the expressions of voltage
are,

$$v(t) = 5t \text{ V} \qquad 0 < t < 1 \tag{2.40}$$

$$\frac{v(t) - 0}{0 - 5} = \frac{t - 2}{2 - 1} \tag{2.41}$$

$$v(t) = -5t + 10 \tag{2.42}$$

$$v(t) = -5t + 10 \text{ V} \qquad 1 < t < 2 \tag{2.43}$$

The average value of the voltage waveform (periodic) is calculated as,

$$V_{av} = \frac{1}{2}\left[\int_0^1 5t\,dt + \int_1^2 (-5t + 10)dt\right] \tag{2.44}$$

$$V_{av} = \frac{2.5}{2}[t^2]_0^1 - \frac{2.5}{2}[t^2]_1^2 + \frac{10}{2}[t]_1^2 \tag{2.45}$$

$$V_{av} = \frac{2.5}{2} - \frac{2.5}{2} \times 3 + 5 = 2.5 \text{ V} \tag{2.46}$$

**Fig. 2.10** Waveform for
practice problem 2.3

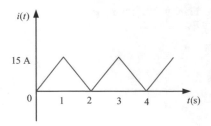

**Practice Problem 2.3**
A current waveform is shown in Fig. 2.10. Determine the average value of this current
waveform.

## 2.5   RMS Value for Complex Waveform

Consider that an instantaneous value $i$ of a periodic (with a time period $T$) complex
current waveform that flows through a resistor $R$ is given by,

$$i = I_0 + I_{m1} \sin(\omega t + \theta_1) + I_{m2} \sin(2\omega t + \theta_2) + I_{m3} \sin(3\omega t + \theta_3) \qquad (2.47)$$

According to Eq. (2.24), the rms values of $I_{m1} \sin(\omega t + \theta_1)$, $I_{m2} \sin(2\omega t + \theta_2)$,
and $I_{m3} \sin(3\omega t + \theta_3)$ are,

$$I_1 = \frac{I_{m1}}{\sqrt{2}} \qquad (2.48)$$

$$I_2 = \frac{I_{m2}}{\sqrt{2}} \qquad (2.49)$$

$$I_3 = \frac{I_{m3}}{\sqrt{2}} \qquad (2.50)$$

Heat produced due to the first component of the current is,

$$H_1 = I_0^2 RT \qquad (2.51)$$

Heat produced due to the second component of the current is,

$$H_2 = \left(\frac{I_{m1}}{\sqrt{2}}\right)^2 RT \qquad (2.52)$$

Heat produced due to the third component of the current is,

$$H_3 = \left(\frac{I_{m2}}{\sqrt{2}}\right)^2 RT \tag{2.53}$$

Total heat produced by these current components can be expressed as,

$$H_t = I_0^2 RT + \left(\frac{I_{m1}}{\sqrt{2}}\right)^2 RT + \left(\frac{I_{m2}}{\sqrt{2}}\right)^2 RT \tag{2.54}$$

According to the definition of rms value, Eq. (2.54) can be modified as,

$$I^2 RT = I_0^2 RT + \left(\frac{I_{m1}}{\sqrt{2}}\right)^2 RT + \left(\frac{I_{m2}}{\sqrt{2}}\right)^2 RT \tag{2.55}$$

$$I_{rms} = \sqrt{I_0^2 + \left(\frac{I_{m1}}{\sqrt{2}}\right)^2 + \left(\frac{I_{m2}}{\sqrt{2}}\right)^2} \tag{2.56}$$

Similarly, for instantaneous value of a complex voltage waveform, can be expressed as,

$$V_{rms} = \sqrt{V_0^2 + \left(\frac{V_{m1}}{\sqrt{2}}\right)^2 + \left(\frac{V_{m2}}{\sqrt{2}}\right)^2} \tag{2.57}$$

**Example 2.4** A complex voltage waveform is given by $v(t) = 2 + 3 \sin \omega t + 5 \sin 2\omega t$ V. Calculate the rms value and the power absorbed by a $2\Omega$ resistor.

**Solution:**
The rms value of the complex voltage waveform can be calculated as,

$$V_{rms} = \sqrt{2^2 + \left(\frac{3}{\sqrt{2}}\right)^2 + \left(\frac{5}{\sqrt{2}}\right)^2} = 4.58 \text{ V} \tag{2.58}$$

The power absorbed by the $2\Omega$ resistor is calculated as,

$$P = \frac{V_{rms}^2}{R} = \frac{4.58^2}{2} = 10.49 \text{ W} \tag{2.59}$$

**Practice Problem 2.4**
A complex voltage waveform is given by $v(t) = 1 + 1.5 \sin \omega t$ V. Determine the rms value and the power absorbed by a $4\Omega$ resistor.

## 2.6   Form Factor and Peak Factor

The ratio of rms value to the average value of an alternating voltage or current is known as the form factor. Mathematically, the form factor is expressed as,

$$\text{Form factor} = \frac{I_{rms}}{I_{av}} = \frac{V_{rms}}{V_{av}} \qquad (2.60)$$

The ratio of maximum value to the rms value of an alternating voltage or current is known as peak factor, which is also known as crest factor. Mathematically, it is expressed as,

$$\text{Peak factor} = \frac{I_m}{I_{rms}} = \frac{V_m}{V_{rms}} \qquad (2.61)$$

## 2.7   Concept of Phasor

Charles Proteus Steinmetz, a mathematician and an electrical engineer established the phasor in an AC circuit between the years 1865–1923. A phasor is a complex term that represents the magnitude and the phase angle of a sinusoidal waveform. Phasor is often known as the vector [5–7]. A definite length of a line rotating in the anticlockwise direction with a constant angular speed is also known as a phasor. As shown in Fig. 2.11, the phasor $P$ rotates in the anticlockwise direction with a constant angular speed $\omega$, where the phase angle $\theta$ is measured with respect to the real axis.

Here, $r$ is the magnitude of the phasor. In a rectangular form, the phasor can be represented as,

$$P = a + jb \qquad (2.62)$$

In a polar form, the phasor is represented as,

**Fig. 2.11**  Phasor with an angular speed

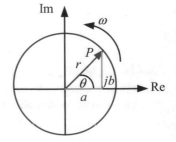

$$P = r\underline{|\theta°}$$ (2.63)

The magnitude and the phase angle of the phasor are calculated as,

$$r = \sqrt{a^2 + b^2}$$ (2.64)

$$\theta = \tan^{-1}\left(\frac{b}{a}\right)$$ (2.65)

Again, from Fig. 2.11, the following expressions relationship can be written:

$$a = r\cos\theta$$ (2.66)

$$b = r\sin\theta$$ (2.67)

Substituting Eqs. (2.66) and (2.67) into Eq. (2.62) yields,

$$P = r\cos\theta + jr\sin\theta = r(\cos\theta + j\sin\theta) = r\underline{|\theta}$$ (2.68)

The phasor in an exponential form is expressed as,

$$P = re^{j\theta}$$ (2.69)

where the following is expressed as,

$$e^{j\theta} = \cos\theta + j\sin\theta$$ (2.70)

A sinusoidal voltage $v(t) = V_m\sin(\omega t + \theta)$ can be expressed as,

$$v(t) = \text{Im}(V_m e^{j(\omega t + \theta)}) = \text{Im}(V_m e^{j\omega t}e^{j\theta}) = \text{Im}(Ve^{j\omega t})$$ (2.71)

where the phasor voltage is expressed as,

$$V = V_m e^{j\theta} = V_m\underline{|\theta}$$ (2.72)

The derivative of a sinusoidal voltage $v(t) = V_m\sin(\omega t + \theta)$ can be expressed as,

$$\frac{dv(t)}{dt} = V_m\omega\cos(\omega t + \theta) = V_m\omega\sin\{90° + (\omega t + \theta)\}$$ (2.73)

$$\frac{dv(t)}{dt} = \omega\text{Im}(V_m e^{j90°}e^{j\omega t}e^{j\theta}) = j\omega\text{Im}(Ve^{j\omega t})$$ (2.74)

$$\frac{dv(t)}{dt} = j\omega V$$ (2.75)

Similarly, the following expression:

$$\int v(t)dt = \frac{V}{j\omega} \qquad (2.76)$$

***Example 2.5*** Convert the voltage $v(t) = 10\sin(10t - 45°)$ V and current $i(t) = 5\cos(20t - 65°)$ A expressions into their phasor forms.

**Solution:**
The phasor form of voltage can be written as,

$$V = 10\underline{|{-45°}}\, \text{V} \qquad (2.77)$$

The current expression can be re-arranged as,

$$i(t) = 5\cos(20t - 65°) = 5\sin(20t - 65° + 90°) = 5\sin(20t + 25°)\,\text{A} \quad (2.78)$$

The phasor form of current can be expressed as,

$$I = 5\underline{|25°}\,\text{A} \qquad (2.79)$$

**Practice Problem 2.5**
The voltage and the current expressions are given by $v(t) = 2\sin(314t + 25°)$ V and $i(t) = -5\sin(-20t - 95°)$ A, respectively. Convert these expressions into phasor forms.

## 2.8   The j-Operator

The $j$-operator ($j = \sqrt{-1}$) is an important multiplication factor in an AC circuit. It provides a 90° displacement in the anticlockwise direction from the reference axis when multiplying with any phasor. Consider a phasor $P$ works in the positive x-axis as shown in Fig. 2.12. Multiply the phasor P by j and the new phasor $jP$ works in the positive y-axis which is 90° apart from the x-axis. Again, multiplying the phasor $jP$ by $j$ results in the new phasor $jjP$ works in the negative x-axis.

In the positive and negative x-axis, the phasors are equal, but work in the opposite directions and in this case, the associated expression can be written as,

$$-jjP = P \qquad (2.80)$$

$$j^2 = -1 \qquad (2.81)$$

**Fig. 2.12** Phasor with a
quadrant

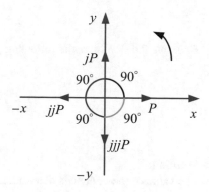

## 2.9 Phasor Algebra

Phasor addition, subtraction, multiplication, division, power and conjugate have been
discussed here. Consider the following phasors to discuss phasor algebra.

$$p_1 = a_1 + jb_1 = r_1 \lfloor \theta_1 \tag{2.82}$$

$$p_2 = a_2 + jb_2 = r_2 \lfloor \theta_2 \tag{2.83}$$

The phasor addition is expressed as,

$$p_a = a_1 + jb_1 + a_2 + jb_2 = (a_1 + a_2) + j(b_1 + b_2) \tag{2.84}$$

The phasor subtraction is expressed as,

$$p_s = a_1 + jb_1 - a_2 - jb_2 = (a_1 - a_2) + j(b_1 - b_2) \tag{2.85}$$

The phasor multiplication is expressed as,

$$p_m = r_1 \lfloor \theta_1 \times r_2 \lfloor \theta_2 = r_1 r_2 \lfloor \theta_1 + \theta_2 \tag{2.86}$$

The phasor division is expressed as,

$$p_d = r_1 \lfloor \theta_1 \div r_2 \lfloor \theta_2 = \frac{r_1}{r_2} \lfloor \theta_1 - \theta_2 \tag{2.87}$$

The power of the phasor is expressed as,

$$p_p = \left( r \lfloor \theta \right)^n = r^n \lfloor \theta \tag{2.88}$$

The conjugate of any phasor is expressed as,

$$p_{conj} = a - jb = r\underline{|-\theta} \tag{2.89}$$

**Example 2.6** Evaluate the following phasor expression in polar form:

$$A = \frac{(2 + j5 + 3 - j2)(3 - j2)}{(2 + j3)^3}$$

**Solution:**
The value of the voltage is determined as,

$$A = \frac{(2 + j5 + 3 - j2)(3 - j2)}{(2 + j3)^3} = \frac{21.02\underline{|-2.73°}}{46.87\underline{|168.93°}} = 0.45\underline{|-171.66°} \tag{2.90}$$

**Practice Problem 2.6**
Evaluate the following expression in polar form:

$$B = \frac{(6 + j5)(3 - j2)}{(5 + j3)(-j3)}$$

## 2.10 Alternating Current with Resistor

Figure 2.13 shows a AC circuit with a resistor. Consider that the expression of the source voltage in this circuit is,

$$v(t) = V_m \sin \omega t \tag{2.91}$$

The phasor form of this voltage is,

$$V = V_m\underline{|0°} \tag{2.92}$$

**Fig. 2.13** Resistance in series with a voltage source

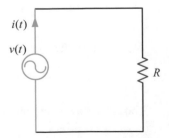

The current in the resistor $R$ is,

$$i(t) = \frac{v(t)}{R} \tag{2.93}$$

Substituting Eq. (2.90) into Eq. (2.92) yields,

$$i(t) = \frac{V_m \sin \omega t}{R} \tag{2.94}$$

$$i(t) = I_m \sin \omega t \tag{2.95}$$

The phasor form of this current is,

$$I = I_m \underline{0°} \tag{2.96}$$

The voltage and current waveforms are drawn in the same graph as shown in Fig. 2.14. The starting point of both the waveforms is the same. Therefore, the voltage and the current waveforms are in the same phase. The impedance (impedance is the resistance to AC circuit that contains both magnitude and phase; resistance contains magnitude only) due to the resistance is written as,

$$Z_R = R + j0 \tag{2.97}$$

The real part (Re) and the imaginary part ($I_m$) of impedance due to resistance are plotted as shown in Fig. 2.15.

***Example 2.7*** A $3\Omega$ resistor is connected in series with a voltage source whose expression is given by $v(t) = 6\sin(314t + 25°)$ V. Calculate the current in the resistor using the phasor, and draw the phasor diagram for the associated voltage and current.

**Solution:**
The phasor form of the voltage is,

**Fig. 2.14** Voltage and current are in phase

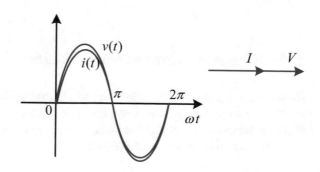

**Fig. 2.15** Phasor diagram
for impedance due to
resistance

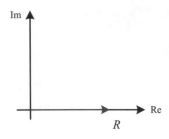

**Fig. 2.16** Phasor diagram
for Example 2.7

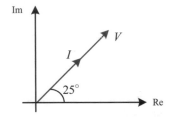

$$V = 6\underline{25°} \, V \tag{2.98}$$

The value of the current in the phasor form is,

$$I = \frac{6\underline{25°}}{3} = 2\underline{25°} \, A \tag{2.99}$$

The current in sinusoidal form is,

$$i(t) = 2\sin(314t + 25°) \, A \tag{2.100}$$

The phasor diagram is shown in Fig. 2.16.

**Practice problem 2.7**
A current $i(t) = 5\sin(377t - 35°)$ A flows through a 5Ω resistor. Use phasor to find
the voltage across the resistor and draw the phasor diagram of the associated current
and voltage.

## 2.11 Alternating Current with Inductor

An inductor is connected in series with an alternating voltage source as shown in
Fig. 2.17. An alternating current will flow through the inductor. As a result, an emf
will be induced in the coil and this induced emf will be equal to the applied voltage.
The induced emf in the inductor is expressed as,

**Fig. 2.17**  Inductor in series
with a voltage source

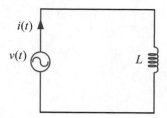

$$v(t) = L\frac{di(t)}{dt} \tag{2.101}$$

Substituting Eq. (2.91) into Eq. (2.101) yields,

$$V_m \sin \omega t = L\frac{di(t)}{dt} \tag{2.102}$$

Rearranging Eq. (2.102) yields,

$$di(t) = \frac{V_m}{L} \sin \omega t \ dt \tag{2.103}$$

Integrating Eq. (2.103) yields,

$$i(t) = -\frac{V_m}{\omega L} \cos \omega t = \frac{V_m}{\omega L} \sin\left(\omega t - \frac{\pi}{2}\right) \tag{2.104}$$

The current in phasor form is,

$$I = I_m \underline{|-90°} \tag{2.105}$$

where, the maximum value of the current is,

$$I_m = \frac{V_m}{\omega L} \tag{2.106}$$

The phase difference between voltage and current is $\phi_{pd} = 0 + 90° = 90°$. Therefore, voltage leads the current by an angle 90° or current lags the voltage waveform by the same angle. The waveforms for the voltage and current are shown in Fig. 2.18.

Impedance due to inductance is defined as the ratio of the phasor form of voltage to the phasor form of current, and it is written as,

$$Z_L = \frac{V}{I} \tag{2.107}$$

Substituting Eqs. (2.92) and (2.105) into Eq. (2.107) yields,

**Fig. 2.18** Voltage and
current waveforms

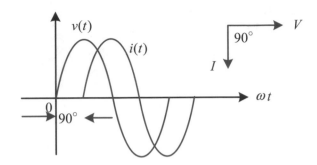

$$Z_L = \frac{V_m\,\underline{|0°}}{\frac{V_m}{\omega L}\,\underline{|-90°}} = \omega L\,\underline{|90°} = j\omega L \qquad (2.108)$$

The magnitude $Z_L$ is called the inductive reactance and it is represented by $X_L$. Equation (2.108) can be modified as,

$$Z_L = jX_L \qquad (2.109)$$

where, the expression of inductive reactance is expressed as,

$$X_L = \omega L \qquad (2.110)$$

The phasor representation of impedance due to inductance is shown in Fig. 2.19.

***Example 2.8*** The voltage across a 2 mH inductor is found to be $v(t) = 5\sin(34t + 35°)$ V. Determine the current in the inductor using the phasor and draw the phasor diagram for the associated current and voltage

**Solution:**
The phasor form of the voltage is,

$$V = 5\underline{|35°}\,\text{V} \qquad (2.111)$$

**Fig. 2.19** Phasor diagram
for impedance due to
inductance

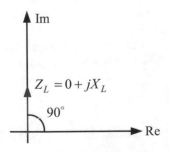

**Fig. 2.20** Phasor diagram
for Example 2.8

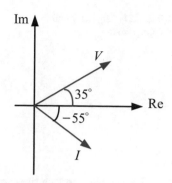

The inductive reactance is,

$$X_L = \omega L = 34 \times 2 \times 10^{-3} = 0.086\,\Omega \qquad (2.112)$$

The current is calculated as,

$$I = \frac{V}{jX_L} = \frac{5\underline{|35°}}{0.086\underline{|90°}} = 58.14\underline{|-55°}\,\text{A} \qquad (2.113)$$

The current in sinusoidal form or in the time domain is,

$$i(t) = 58.14\sin(34\,t - 55°)\,\text{A} \qquad (2.114)$$

The phasor diagram is shown in Fig. 2.20.

**Practice problem 2.8**
The voltage across a 5H inductor is measured as $v(t) = 15\sin(30t - 20°)$ V. Calculate the current in the inductor using the phasor, and draw the phasor diagram for the associated current and voltage.

## 2.12 Alternating Current with Capacitor

Figure 2.21 shows a circuit where a capacitor with $C$ is connected in series with an alternating voltage source $v$. In this case, when a current flows through it, the instantaneous charge stored in the capacitor will be,

$$q = Cv \qquad (2.115)$$

Substituting Eq. (2.91) into Eq. (2.115) yields,

$$q = CV_m \sin \omega t \qquad (2.116)$$

**Fig. 2.21** Capacitor with a
voltage source

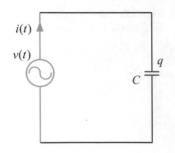

The current through the capacitor is expressed as,

$$i(t) = \frac{dq}{dt} \qquad (2.117)$$

Substituting Eq. (2.116) into Eq. (2.117) yields,

$$i(t) = \frac{d}{dt}(CV_m \sin \omega t) \qquad (2.118)$$

$$i(t) = CV_m \omega \cos \omega t \qquad (2.119)$$

$$i(t) = CV_m \omega \sin(90° + \omega t) \qquad (2.120)$$

where the maximum value of the current is,

$$I_m = C\omega V_m \qquad (2.121)$$

From Eq. (2.120), the expression of the current in phasor form can be written as,

$$I = I_m \underline{|90°} \qquad (2.122)$$

In this case, the phase difference between the voltage and the current is,

$$\theta_{pd} = 0 - 90° = -90° \qquad (2.123)$$

From Eq. (2.123), it is observed that the current waveform in a capacitor leads to the voltage waveform or the voltage waveform lags the current waveform. The voltage and the current waveforms in a capacitor are shown in Fig. 2.22.

The impedance due to capacitance is defined as the ratio of the phasor voltage and phasor current. Mathematically, it is expressed as,

$$Z_C = \frac{V}{I} \qquad (2.124)$$

**Fig. 2.22** Waveforms with lagging phasor diagram

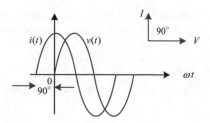

Substituting Eqs. (2.92) and (2.122) into Eq. (2.124) yields,

$$Z_C = \frac{V_m \underline{|0°}}{I_m \underline{|90°}} \tag{2.125}$$

Substituting Eq. (2.121) into Eq. (2.125) yields,

$$Z_C = \frac{V_m \underline{|0°}}{C V_m \omega \underline{|90°}} = \frac{1}{\omega C} \underline{|-90°} \tag{2.126}$$

$$Z_C = \frac{1}{\omega C} \underline{|-90°} = -j \frac{1}{\omega C} \tag{2.127}$$

The magnitude $Z_C$ is called the capacitive reactance and it is represented by $X_C$. Equation (2.127) can be rearranged as,

$$Z_C = -j X_C \tag{2.128}$$

where, the expression of capacitive reactance can be written as,

$$X_C = \frac{1}{\omega C} \tag{2.129}$$

The phasor representation of impedance due to capacitive reactance is shown in Fig. 2.23.

**Fig. 2.23** Phasor diagram for impedance due to capacitance

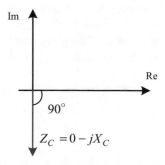

**Example 2.9** Determine the current in both the phasor and sinusoidal forms when a voltage across a 5μF capacitor is given by $v(t) = 65 \sin(700t - 40°)$ V.

**Solution:**
The capacitive reactance is calculated as,

$$X_C = \frac{1}{\omega C} = \frac{1}{700 \times 5 \times 10^{-6}} = 285.71 \, \Omega \tag{2.130}$$

The voltage in phasor form is,

$$V = 65\underline{|-40°} \, \text{V} \tag{2.131}$$

The current in phasor form can be determined as,

$$I = \frac{V}{-jX_C} = \frac{65\underline{|-40°}}{285.71\underline{|-90°}} = 0.23\underline{|50°} \, \text{A} \tag{2.132}$$

The current in sinusoidal form is,

$$i(t) = 0.23 \sin(700t + 50°) \, \text{A} \tag{2.133}$$

**Practice Problem 2.9**
A 3μF capacitor is connected to a voltage source, and the current through this capacitor is given by $i(t) = 12 \sin(377t + 26°)$ A. Determine the voltage across the capacitor in both phasor and sinusoidal forms.

## 2.13   Impedance and Admittance

The impedance and admittance are considered to be important parameters in analyzing AC circuits. The impedance is the characteristic of a circuit or circuit element force that opposes the current flow in an AC circuit. Impedance is equivalent to resistance in a DC circuit. The impedance, which is a vector quality, is also defined as the ratio of the phasor voltage to the phasor current. The impedance is represented by the letter Z, and it is frequency-dependent. The reciprocal of impedance is known as admittance. It is represented by the letter Y and its unit is Siemens (S).

The impedance can be written as,

$$Z = \frac{V}{I} \tag{2.134}$$

where, the admittance can be written as,

**Fig. 2.24** Impedance triangle

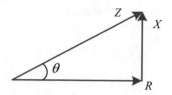

$$Y = \frac{I}{V} = \frac{1}{Z} \tag{2.135}$$

Both impedance and admittance are related to resistance and reactance. The impedance triangle is shown in Fig. 2.24.

According to Fig. 2.24, the expression of impedance can be written as,

$$Z = R + jX \tag{2.136}$$

The impedance in polar form can be written as,

$$Z = |Z|\underline{\theta} \tag{2.137}$$

where the magnitude of the impedance is,

$$|Z| = \sqrt{R^2 + X^2} \tag{2.138}$$

The expression of phase angle is,

$$\theta = \tan^{-1}\left(\frac{X}{R}\right) \tag{2.139}$$

In addition, from Fig. 2.24, the following expressions can be written,

$$R = Z \cos\theta \tag{2.140}$$

$$X = Z \sin\theta \tag{2.141}$$

Substituting Eq. (2.136) into Eq. (2.135) yields,

$$Y = \frac{1}{R + jX} \tag{2.142}$$

Multiplying the numerator and denominator of the right-hand side of Eq. (2.142) by $(R - jX)$ yields,

**Fig. 2.25** Circuit for
Example 2.10

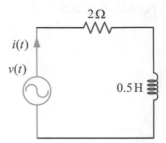

$$Y = \frac{R - jX}{(R + jX)(R - jX)} = \frac{R - jX}{R^2 + X^2} \qquad (2.143)$$

$$Y = \frac{R}{R^2 + X^2} - j\frac{X}{R^2 + X^2} \qquad (2.144)$$

$$Y = G + jB \qquad (2.145)$$

where, $G$ and $B$ are known as the conductance and the susceptance, respectively, and
their units are Siemens (S). The expression of these two parameters are written as,

$$G = \frac{R}{R^2 + X^2} \qquad (2.146)$$

$$B = -\frac{X}{R^2 + X^2} \qquad (2.147)$$

**Example 2.10** Determine the inductive reactance, and the current in both phasor
and sinusoidal forms for the circuit in Fig. 2.25, if the applied voltage is $v(t) = 15\sin(10t - 20°)$ V.

**Solution:**
The value of the inductive reactance is calculated as,

$$X_L = \omega L = 10 \times 0.5 = 5\,\Omega \qquad (2.148)$$

The impedance of the circuit is,

$$Z = 2 + j5 = 3.61 \underline{|56.32°}\ \Omega \qquad (2.149)$$

The voltage in phasor form is,

$$V = 15 \underline{|-20°}\ V \qquad (2.150)$$

The current in the phasor form is calculated as,

$$I = \frac{V}{Z} = \frac{15\underline{|-20°}}{3.61\underline{|56.31°}} = 4.16\underline{|-76.31°}\,\text{A} \tag{2.151}$$

The expression of the current in the sinusoidal form is,

$$i(t) = 4.16\sin(10t - 76.31°)\,\text{A} \tag{2.152}$$

***Example 2.11*** Two circuit components with impedances $Z_1 = 2 + j3\,\Omega$ and $Z_2 = 4 - j7\,\Omega$ are connected in parallel to an $V = 220\underline{|35°}\,V$ AC source. Calculate the total admittance and the current in both phasor and sinusoidal forms.

**Solution:**
The total admittance is calculated as,

$$Y = Y_1 + Y_2 = \frac{1}{Z_1} + \frac{1}{Z_2} = \frac{1}{2 + j3} + \frac{1}{4 - j7} = 0.25\underline{|-29.74°}\,\text{S} \tag{2.153}$$

The current is calculated as,

$$I = VY = 220\underline{|35°} \times 0.25\underline{|-29.74°} = 55\underline{|-5.26°}\,\text{A} \tag{2.154}$$

**Practice Problem 2.10**
Figure 2.26 shows an RC series circuit. Calculate the capacitive reactance, and the current in both phasor and sinusoidal forms, if the applied voltage is $v(t) = 5\sin(20t + 35°)\,\text{V}$.

**Practice Problem 2.11**
Find the total admittance and the current in both phasor and sinusoidal forms when two circuit components with impedances $Z_1 = 3 + j5\,\Omega$ and $Z_2 = 3 - j4\,\Omega$ are connected in parallel to an $V = 230\underline{|20°}\,\text{V}$ AC source.

**Fig. 2.26** Circuit for practice problem 2.10

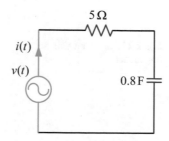

## 2.14  Instantaneous Power

Figure 2.27 shows a circuit with an AC voltage source $v(t)$ with an impedance connected in series. In this case, the resultant current $i(t)$ varies with time. The instantaneous power also varies with time, and it is defined as the product of the voltage and the current. Let us consider that the expression of voltage is,

$$v(t) = V_m \sin \omega t \tag{2.155}$$

The expression of the current can be derived as [1, 2],

$$i(t) = \frac{v(t)}{Z\underline{|\theta}} \tag{2.156}$$

Substituting Eq. (2.155) into Eq. (2.156) yields,

$$i(t) = \frac{V_m \sin \omega t}{Z\underline{|\theta}} \tag{2.157}$$

$$i(t) = \frac{V_m \underline{|0^\circ}}{Z\underline{|\theta}} = \frac{V_m}{Z}\underline{|-\theta} \tag{2.158}$$

$$i(t) = I_m\underline{|-\theta} = I_m \sin(\omega t - \theta) \tag{2.159}$$

The expression of instantaneous power is,

$$p(t) = v(t) \times i(t) \tag{2.160}$$

Substituting Eqs. (2.155) and (2.159) into Eq. (2.160) yields,

$$p(t) = V_m \sin \omega t \times I_m \sin(\omega t - \theta) \tag{2.161}$$

$$p(t) = \frac{V_m I_m}{2} 2 \sin \omega t \times \sin(\omega t - \theta) \tag{2.162}$$

**Fig. 2.27**  A circuit with an impedance

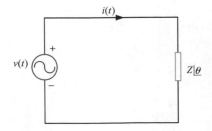

$$p(t) = \frac{V_m I_m}{2}[\cos(\omega t - \omega t + \theta) - \cos(\omega t + \omega t - \theta)] \qquad (2.163)$$

$$p(t) = \frac{V_m I_m}{2}\cos\theta - \frac{V_m I_m}{2}\cos(2\omega t - \theta) \qquad (2.164)$$

Equation (2.164) provides the expression for the instantaneous power for a series AC circuit.

***Example 2.12*** The excitation voltage and impedance of a series circuit are given by $v(t) = 15\sin\omega t$ V and $Z = 5\underline{/10°}$ $\Omega$, respectively. Calculate the instantaneous power.

**Solution**
The value of the current is calculated as,

$$i(t) = \frac{15\underline{/0°}}{5\underline{/10°}} = 3\underline{/-10°} \text{ A} \qquad (2.165)$$

The instantaneous power is calculated as,

$$p(t) = \frac{15 \times 3}{2}2\sin\omega t \times \sin(\omega t - 10°) \qquad (2.166)$$

$$p(t) = \frac{45}{2}\cos 10° - \frac{45}{2}\cos(2\omega t - 10°) \qquad (2.167)$$

$$p(t) = 22.16 - 22.5\cos(2\omega t - 10°) \qquad (2.168)$$

**Practice Problem 2.12**
The current and impedance of a series circuit are given by $i(t) = 10\sin(\omega t + 15°)$ A and $Z = 2\underline{/20°}$ $\Omega$, respectively. Calculate the instantaneous voltage and power.

## 2.15   Average Power and Reactive Power

The average power is related to the sinusoidal voltage and current, which are shown in Eqs. (2.160) and (2.164), respectively. The average power for a periodic waveform over one cycle can be derived as [3, 4],

$$P = \frac{1}{T}\int_0^T p(t) \qquad (2.169)$$

Substituting Eq. (2.164) into the Eq. (2.169) yields,

$$P = \frac{V_m I_m}{2T} \int_0^T \cos\theta \, dt - \frac{V_m I_m}{2T} \int_0^T \cos(2\omega t - \theta) \, dt \qquad (2.170)$$

$$P = \frac{V_m I_m}{2T} \cos\theta [T] - \frac{V_m I_m}{2T} \int_0^T \cos(2\omega t - \theta) \, dt \qquad (2.171)$$

The second term of Eq. (2.171) is a cosine waveform. The average value of any cosine waveform over one cycle is zero. Therefore, from Eq. (2.171), the final expression of the average power can be represented as,

$$P = \frac{V_m I_m}{2} \cos\theta \qquad (2.172)$$

Similarly, the expression of the average reactive power can be written as,

$$Q = \frac{V_m I_m}{2} \sin\theta \qquad (2.173)$$

The term $\cos\theta$ in Eq. (2.1.18) is the power factor of the circuit, and it is determined by the phase angle $\theta$ of the circuit impedance, where $\theta$ is the phase difference between the voltage and current phases i. e., $\theta = \theta_v - \theta_i$. The average power is often known as the true power or real power. The units of average power and reactive power are watts (W) and volt-ampere reactive (Var), respectively. The average power from Eq. (2.172) can be represented in terms of rms values of the voltage and current as,

$$P = \frac{V_m}{\sqrt{2}} \frac{I_m}{\sqrt{2}} \cos\theta = \frac{V_m}{\sqrt{2}} \frac{I_m}{\sqrt{2}} \cos(\theta_v - \theta_i) \qquad (2.174)$$

Substituting $V_{rms} = \frac{V_m}{\sqrt{2}}$ and $I_{rms} = \frac{I_m}{\sqrt{2}}$ into Eq. (2.174) yields,

$$P = V_{rms} I_{rms} \cos\theta = V_{rms} I_{rms} \cos(\theta_v - \theta_i) \qquad (2.175)$$

Similarly, from Eq. (2.173), the reactive power can be represented as,

$$Q = V_{rms} I_{rms} \sin\theta = V_{rms} I_{rms} \sin(\theta_v - \theta_i) \qquad (2.176)$$

Due to a sufficient magnitude of the reactive power, the current flows back and forth between the source and the network. The reactive power does not dissipate any energy in the load. However, in practice, it produces energy losses in the line. Therefore, extra care needs to be taken in designing a power system network.

For a purely resistive circuit, the voltage ($V = V_m \underline{|\theta_v}$) and the current ($I = I_m \underline{|\theta_i}$) are in phase. It means that the phase angle between them is zero,

$$\theta = \theta_v - \theta_i = 0 \tag{2.177}$$

Substituting Eq. (2.177) into Eq. (2.174) yields,

$$P_R = \frac{V_m I_m}{2} \cos 0° \tag{2.178}$$

$$P_R = \frac{V_m I_m}{2} \tag{2.179}$$

Equation (2.179) can be rearranged as,

$$P_R = \frac{I_m^2 R}{2} = \frac{V_m^2}{2R} \tag{2.180}$$

The phase difference between the voltage and current due to inductance and capacitance is,

$$\theta = \theta_v - \theta_i = \pm 90° \tag{2.181}$$

Substituting Eq. (2.181) into Eq. (2.172) yields the average power for either an inductance or a capacitance,

$$P_L = P_C = \frac{V_m I_m}{2} \cos 90° = 0 \tag{2.182}$$

The reactive power is usually stored in a circuit and it can be expressed for the inductor and capacitor as,

$$Q_L = I_L^2 X_L = \frac{V_L^2}{X_L} \tag{2.183}$$

$$Q_C = I_C^2 X_C = \frac{V_C^2}{X_C} \tag{2.184}$$

From Eqs. (2.180) and (2.182), it can be concluded that the resistive load absorbs power whereas inductive or capacitive loads do not absorb any power.

**Example 2.13** An electrical series circuit with resistance, inductive and capacitive reactance is shown in Fig. 2.28. Calculate the average power supplied by the source and the power absorbed by the resistor.

**Solution**
The net impedance is calculated as,

**Fig. 2.28** Circuit for
Example 2.13

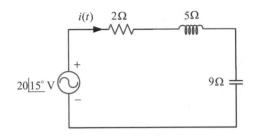

$$Z_t = 2 + j5 - j9 = 4.47 \underline{|-63.43°}\ \Omega \tag{2.185}$$

The source current in phasor form is calculated as,

$$I = \frac{20 \underline{|15°}}{4.47 \underline{|-63.43°}} = 4.47 \underline{|78.43°}\ \text{A} \tag{2.186}$$

The average power supplied by the source is calculated as,

$$P_s = \frac{20 \times 4.47}{2} \cos(15° - 78.43°) = 20\,\text{W} \tag{2.187}$$

The average power absorbed by the resistor is calculated as,

$$P_R = \frac{4.47^2 \times 2}{2} = 20\,\text{W} \tag{2.188}$$

**Practice Problem 2.13**

A series-parallel circuit with resistance, inductive and capacitive reactance is shown
in Fig. 2.29. Determine the average power supplied by the source and the power
absorbed by the resistors.

**Fig. 2.29** Circuit for
practice problem 2.13

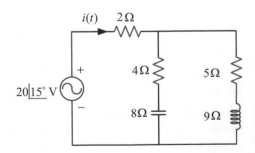

## 2.16  Apparent Power

The apparent power can be derived from the average power. But, the apparent power is related to the sinusoidal voltage and current. Let us consider that the expressions for sinusoidal voltage and current are [5, 6],

$$v(t) = V_m \sin(\omega t + \theta_v) \tag{2.189}$$

$$i(t) = I_m \sin(\omega t + \theta_i) \tag{2.190}$$

The phasor forms of these voltage and current components are,

$$V = V_m | \theta_v \tag{2.191}$$

$$I = I_m \underline{| \theta_i} \tag{2.192}$$

According to Eq. (2.174), the average power is,

$$P = \frac{V_m I_m}{2} \cos(\theta_v - \theta_i) = V_{rms} I_{rms} \cos(\theta_v - \theta_i) \tag{2.193}$$

The apparent power is the product of the rms voltage and rms current. The unit of apparent power is Volt-Amps (VA) and is denoted by the letter $S$. The apparent power can be expressed as,

$$S = \frac{V_m I_m}{2} = V_{rms} I_{rms} \tag{2.194}$$

Substituting Eq. (2.194) into Eq. (2.193) yields,

$$P = S \cos(\theta_v - \theta_i) \tag{2.195}$$

In addition, the apparent power can be determined by the vector sum of the real power ($P$) and the reactive power ($Q$). In this case, the expression of reactive power becomes,

$$S = P + jQ \tag{2.196}$$

The power triangle (Power Triangle): A right-angle triangle that shows the vector relationship between active power, reactive power and apparent power) with a lagging and leading power factor is shown in Fig. 2.30. The power triangles with an inductance and the capacitance loads will be lagging and leading, respectively.

**Fig. 2.30** Power triangles

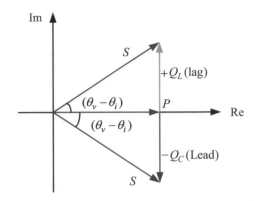

*Example 2.14* An industrial load draws a current $i(t) = 16\sin(314t + 25°)$ A from an alternating voltage source $v(t) = 220\sin(314t + 60°)$ V. Determine the apparent power, circuit resistance and inductance.


**Solution**

The apparent power is calculated as,

$$S = \frac{V_m I_m}{2} = \frac{220 \times 16}{2} = 1.76\,\text{kVA} \tag{2.197}$$

The circuit impedance is calculated as,

$$Z = \frac{220\underline{|60°}}{16\underline{|25°}} = 11.26 + j7.89\,\Omega \tag{2.198}$$

The value of the circuit resistance is,

$$R = 11.26\,\Omega \tag{2.199}$$

The circuit inductance is calculated as,

$$L = \frac{7.89}{314} = 0.025\,\text{H} \tag{2.200}$$


**Practice Problem 2.14**

An industrial load draws a current $i(t) = 10\sin(100t + 55°)$ A from an alternating voltage source $v(t) = 120\sin(100t + 10°)$ V. Calculate the apparent power, circuit resistance and capacitance.

## 2.17 Complex Power

Complex power is the combination of real power and reactive power. The reactive power creates an adverse effect on power generation, which can be studied by analyzing the complex power. Mathematically, the product of half of the phasor voltage and the conjugate of the phasor current is known as complex power. The complex power is represented by the letter $S_c$ and is expressed as,

$$S_c = \frac{1}{2} V I^*$$ (2.201)

Substituting Eqs. (2.191) and (2.192) into Eq. (2.201) yields,

$$S_c = \frac{1}{2} V_m \underline{|\theta_v} \times I_m \underline{|-\theta_i}$$ (2.202)

$$S_c = V_{rms} \underline{|\theta_v} \times I_{rms} \underline{|-\theta_i}$$ (2.203)

The complex power in terms of phasor form of rms voltage and current can be written as,

$$S_c = \mathbf{V}_{rms} \times \mathbf{I}^*_{rms}$$ (2.204)

From Eq. (2.204), the complex power is defined as the product of rms voltage and the conjugate of the rms current.

Equation (2.203) can be re-arranged as,

$$S_c = V_{rms} I_{rms} \underline{|\theta_v - \theta_i}$$ (2.205)

$$S_c = V_{rms} I_{rms} \cos(\theta_v - \theta_i) + j V_{rms} I_{rms} \sin(\theta_v - \theta_i)$$ (2.206)

**Fig. 2.31** A simple AC circuit

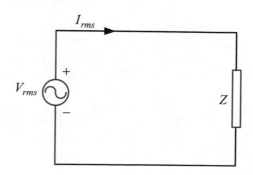

Consider the circuit as shown in Fig. 2.31 to explain the complex power. The impedance of this circuit is,

$$Z = R + jX \tag{2.207}$$

The rms value of the current is,

$$I_{rms} = \frac{V_{rms}}{Z} \tag{2.208}$$

Substituting Eq. (2.208) into Eq. (2.204) yields,

$$S = V_{rms} \frac{V_{rms}^*}{Z^*} = \frac{V_{rms}^2}{Z} \tag{2.209}$$

Equation (2.204) again can be represented as,

$$S_c = I_{rms} Z I_{rms}^* = I_{rms}^2 Z \tag{2.210}$$

Substituting Eq. (2.207) into Eq. (2.210) yields,

$$S_c = I_{rms}^2 (R + jX) = I_{rms}^2 R + j I_{rms}^2 X = P + jQ \tag{2.211}$$

where, $P$ and $Q$ are the real and the imaginary parts of the complex power, and in this case, the expressions of $P$ and $Q$ can be written as,

$$P = \text{Re}(S_c) = I_{rms}^2 R \tag{2.212}$$

$$Q = \text{Im}(S_c) = I_{rms}^2 X \tag{2.213}$$

The real power, reactive power and apparent power of Eq. (2.211) are shown in Fig. 2.32. The complex power for a resistive branch can be written as,

$$S_{cR} = P_R + jQ_R = I_{rms}^2 R \tag{2.214}$$

**Fig. 2.32** Power triangle

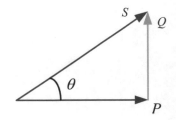

From Eq. (2.214), the real power and the reactive power for the resistive branch can be expressed as,

$$P_R = I_{rms}^2 R \tag{2.215}$$

$$Q = 0 \tag{2.216}$$

The complex power for an inductive branch is,

$$S_{cL} = P_L + jQ_L = jI_{rms}^2 X_L \tag{2.217}$$

From Eq. (2.217), the real power and the reactive power for an inductive branch can be separated as,

$$P_L = 0 \tag{2.218}$$

$$Q_L = I_{rms}^2 X_L \tag{2.219}$$

The complex power for a capacitive branch is,

$$S_{cC} = P_C + jQ_C = -jI_{rms}^2 X_C \tag{2.220}$$

From Eq. (2.220), the real power and the reactive power for a capacitive branch can be separated as,

$$P_c = 0 \tag{2.221}$$

$$Q_C = I_{rms}^2 X_C \tag{2.222}$$

In case of reactive power, the following points are summarized:
$Q = 0$ for resistive load, i.e., unity power factor,
$Q > 0$ for inductive load, i.e., lagging power factor,
$Q < 0$ for capacitive load, i.e., leading power factor.

**Example 2.15** A series AC circuit is shown in Fig. 2.33. Calculate the source current, apparent, real and reactive powers. The expression of the alternating voltage source is $v(t) = 16\sin(10t + 25°)$ V.

**Solution**
The rms value of the source voltage is calculated as,

$$V_{rms} = \sqrt{2} \times 16\underline{|25°} = 22.63\underline{|25°} \text{ V} \tag{2.223}$$

**Fig. 2.33** Circuit for
Example 2.15

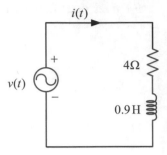

The value of the inductive reactance is,

$$X_L = 10 \times 0.09 = 9\,\Omega \tag{2.224}$$

The circuit impedance is calculated as,

$$Z = 4 + j9 = 9.85\underline{|66.04°}\,\Omega \tag{2.225}$$

The source current is calculated as,

$$I_{rms} = \frac{22.63\underline{|25°}}{9.85\underline{|66.04°}} = 2.30\underline{|-41.04°}\,\text{A} \tag{2.226}$$

The complex power is calculated as,

$$S = V_{rms}I^*_{rms} = 22.63\underline{|25°} \times 2.30\underline{|-41.04°} = 50.02\,\text{W} - j14.38\,\text{Var} \tag{2.227}$$

The apparent power is determined as,

$$S_{ap} = |S| = 52.05\,\text{VA} \tag{2.228}$$

The real power is calculated as,

$$P = \text{Re}(S) = 52.02\,\text{W} \tag{2.229}$$

The reactive power is determined as,

$$Q = \text{Im}(S) = 14.38\,\text{Var} \tag{2.230}$$

**Example 2.16** A 220 V rms delivers power to a load. The load absorbs an average power of 10 kW at a leading power factor of 0.9. Determine the complex power and the impedance of the load.

**Solution**

The power factor is,

$$\cos \theta = 0.9 \tag{2.231}$$

$$\theta = 25.84° \tag{2.232}$$

The reactive component is calculated as,

$$\sin \theta = \sin 25.84° = 0.44 \tag{2.233}$$

The magnitude of the complex power is calculated as,

$$|S| = \frac{P}{\cos \theta} = \frac{10}{0.9} = 11.11\,\text{kVA} \tag{2.234}$$

The reactive power is determined as,

$$Q = |S| \sin \theta = 11.11 \times 0.44 = 4.89\,\text{kVar} \tag{2.235}$$

The complex power is calculated as,

$$S = P + jQ = 10\,\text{kW} - j4.89\,\text{kVar} \tag{2.236}$$

The rms voltage can be determined as,

$$P = V_{rms} I_{rms} \cos \theta = 10000 \tag{2.237}$$

$$I_{rms} = \frac{10000}{220 \times 0.9} = 50.51\,\text{A} \tag{2.238}$$

The value of the impedance is calculated as,

$$|Z| = \frac{|V_{rms}|}{|I_{rms}|} = \frac{220}{50.51} = 4.36\,\Omega \tag{2.239}$$

$$Z = 4.36\underline{|-25.84°}\,\Omega \tag{2.240}$$

**Practice Problem 2.15**

A series RC circuit is shown in Fig. 2.34. Find the source current, apparent, real and reactive powers. The expression of an alternating voltage source is $v(t) = 10 \sin(2t + 12°)$ V.

**Fig. 2.34** Circuit for
practice problem 2.15

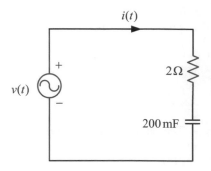

**Practice Problem 2.16**
An electrical load absorbs an average power of 12 kW from a source of 230 V rms
at a lagging power factor of 0.95. Calculate the complex power and the impedance
of the load.

## 2.18  Complex Power Balance

Two electrical loads are connected in parallel with a voltage source as shown in
Fig. 2.35. According to the conservation of energy, the real power delivered by the
source will be equal to the total real power absorbed by the loads [7, 8]. Similarly,
the complex power delivered by the source will be equal to the total complex power
absorbed by the loads. According to KCL, the rms value of the source current is
equal to the sum of the rms values of the branch currents $I_1$ and $I_2$, i.e.,

$$I = I_1 + I_2 \tag{2.241}$$

The total complex power is defined as the product of the rms value of the source
voltage and the conjugate of the current supplied by the source, and it is expressed
as,

**Fig. 2.35** Circuit with two
parallel impedances

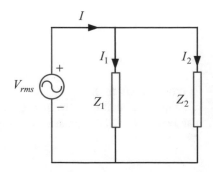

**Fig. 2.36** Circuit with series impedances

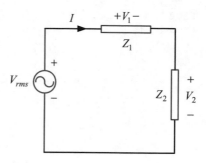

$$S = V_{rms} I^* \tag{2.242}$$

Substituting Eq. (2.241) into Eq. (2.242) yields complex power of the parallel circuit,

$$S_p = V_{rms}[I_1 + I_2]^* \tag{2.243}$$

$$S_p = V_{rms} I_1^* + V_{rms} I_2^* \tag{2.244}$$

$$S_p = S_1 + S_2 \tag{2.245}$$

The electrical loads are again connected in series with a voltage source as shown in Fig. 2.36. According to KVL, the rms value of the source voltage is equal to the sum of the rms values of the load voltages and it is written as,

$$V_{rms} = V_1 + V_2 \tag{2.246}$$

Substituting Eq. (2.246) into Eq. (2.242) yields the complex power of the series circuit is,

$$S_s = (V_1 + V_2)I^* \tag{2.247}$$

$$S_s = V_1 I^* + V_2 I^* \tag{2.248}$$

$$S_s = S_1 + S_2 \tag{2.249}$$

From Eqs. (2.246) and (2.249), it is observed that the total complex power delivered by the source is equal to the sum of the individual complex power, absorbed by the loads.

**Example 2.17** A series-parallel circuit is supplied by a source of 60 V rms as shown in Fig. 2.37. Find the complex power for each branch and the total complex power.

**Fig. 2.37** Circuit for
Example 2.17

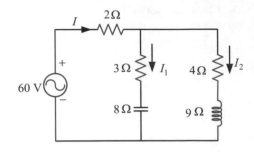

**Solution**

The circuit impedance is calculated as,

$$Z_t = 2 + \frac{(4 + j9)(3 - j8)}{4 + j9 + 3 - j8} = 13.87\underline{|-9.88°}\ \Omega \qquad (2.250)$$

The source current is determined as,

$$I = \frac{60}{13.87\underline{|-9.88°}} = 4.33\underline{|9.88°}\ A \qquad (2.251)$$

The branch currents can be calculated as,

$$I_1 = 4.33\underline{|9.88°} \times \frac{4 + j9}{7 + j1} = 6.03\underline{|67.79°}\ A \qquad (2.252)$$

$$I_2 = 4.33\underline{|9.88°} \times \frac{3 - j8}{7 + j1} = 5.23\underline{|-67.69°}\ A \qquad (2.253)$$

The voltage across the parallel branches is,

$$V_p = 60 - 2 \times 4.33\underline{|9.88°} = 51.49\underline{|-1.65°}\ V \qquad (2.254)$$

The complex power in the branches are calculated as,

$$S_1 = 51.49\underline{|-1.65°} \times 6.03\underline{|-67.79°} = 310.48\underline{|-69.44°}\ VA \qquad (2.255)$$

$$S_2 = 51.49\underline{|-1.65°} \times 5.23\underline{|67.69°} = 269.29\underline{|66.04°}\ VA \qquad (2.256)$$

The voltage drops across the $2\Omega$ resistor is calculated as,

$$V_{2\Omega} = 2 \times 4.33\underline{|9.88°} = 8.66\underline{|9.88°}\ V \qquad (2.257)$$

The complex power for $2\Omega$ resistor is calculated as,

**Fig. 2.38** Circuit for
practice problem 2.17

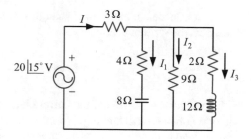

$$S_{2\Omega} = 8.66\underline{9.88°} \times 4.33\underline{-9.88°} = 37.50 \text{ VA} \qquad (2.258)$$

The total complex power is calculated as,

$$S_t = 310.48\underline{-69.44°} + 269.29\underline{66.04°} + 37.50 = 259.76\underline{9.89°} \text{ VA} \qquad (2.259)$$

Alternatively, the total complex power can be calculated as,

$$S = 60 \times 4.33\underline{9.88°} = 259.8\underline{-9.88°} \text{ VA} \qquad (2.260)$$

**Practice Problem 2.17**
A series-parallel circuit is supplied by an rms voltage source as shown in Fig. 2.38.
Determine the complex power for each branch and the total complex power.

## 2.19 Power Factor and Reactive Power

In an AC circuit, power is calculated by multiplying a factor with the rms values of
current and voltage. This factor is known as the power factor. The power factor is
defined as the cosine of the difference in phase angles between the voltage and the
current. Whereas the reactive factor is defined as the sine of the difference in the
phase angles between the voltage and the current. The power factor is also defined
as the cosine of the phase angle of the load impedance. A close to unity power
factor represents an efficient power transfer from the source to a load, whereas a
low power factor identifies an inefficient transmission of power. Low power factor
usually affects power generation devices. Mathematically, the power factor is written
as [9],

$$pf = \cos(\theta_v - \theta_i) \qquad (2.261)$$

The reactive factor is written as,

$$rf = \sin(\theta_v - \theta_i) \qquad (2.262)$$

From impedance and power triangles as shown in Figs. 2.6 and 2.7, the power factor can be written as,

$$\text{pf} = \cos\theta = \frac{R}{Z} = \frac{\text{kW}}{\text{kVA}} \tag{2.263}$$

The angle $\theta$ is positive if the current lags the voltage, and in this case, the power factor is considered as lagging. Whereas the angle $\theta$ is negative if the current leads the voltage, and in this case, the power factor is considered as leading. The leading power factor is usually considered for capacitive loads. The industrial loads are inductive and have a low lagging power factor.

A low power factor has many disadvantages, which are outlined below:

(i)    kVA rating of electrical machines is increased,
(ii)   larger conductor size is required to transmit or distribute electric power at a constant voltage,
(iii)  copper losses are increased, and
(iv)   voltage regulation is small.

## 2.20   Power Factor Correction

In the electrical domain, heavy and medium-sized industry applications contain inductive loads which draw a lagging current from the source. As a result, the reactive power for these applications is increased. In this scenario, the transformer rating and the conductor size need to be increased to carry out the additional reactive power.

In order to cancel this reactive component of power, an opposite type of reactance need to be included in the circuit. Let us consider that a single-phase inductive load is connected across a voltage source as shown in Fig. 2.39 and this load draws a current with a lagging power factor of $\cos\theta_1$.

Figure 2.40 shows a circuit where the capacitor is connected in parallel with the load to improve the power factor. The capacitor will draw current from the source that leads the source voltage by 90°. The line current is the vector sum of the currents in the inductive load and the capacitor. The current in the inductive load circuit lags

**Fig. 2.39** A single-phase inductive circuit

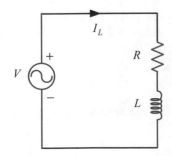

**Fig. 2.40** A Capacitor is in parallel with inductive load

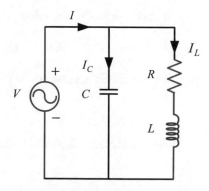

the supply voltage by $\theta_1$, and the current in the capacitor leads the voltage by 90° as shown in the vector diagram in Fig. 2.41.

The exact value of the capacitor needs to be identified to improve the power factor from $\cos\phi_1$ to $\cos\phi_2$ without changing the real power. A power triangle is drawn using the inductive load and the capacitor as shown in Fig. 2.42.

The reactive power of the original inductive load is written as,

$$Q_1 = P \tan \phi_1 \qquad (2.264)$$

**Fig. 2.41** A vector diagram with different currents

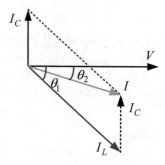

**Fig. 2.42** Power triangles for inductive load and capacitor

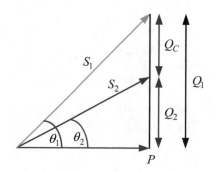

The expression of new reactive power is written as,

$$Q_2 = P \tan \phi_2 \tag{2.265}$$

The reduction in reactive power due to parallel capacitor is expressed as,

$$Q_C = Q_1 - Q_2 \tag{2.266}$$

Substituting Eqs. (2.264) and (2.265) into Eq. (2.266) yields,

$$Q_C = P(\tan \phi_1 - \tan \phi_2) \tag{2.267}$$

The reactive power due to the capacitor can be calculated as,

$$Q_C = \frac{V_{rms}^2}{X_C} = \omega C V_{rms}^2 \tag{2.268}$$

Substituting Eq. (2.267) into Eq. (2.268) yields the expression for the capacitor,

$$\omega C V_{rms}^2 = P(\tan \phi_1 - \tan \phi_2) \tag{2.269}$$

$$C = \frac{P(\tan \phi_1 - \tan \phi_2)}{\omega V_{rms}^2} \tag{2.270}$$

***Example 2.18***  A load of 6 kVA, 50 Hz, 0.75 lagging power factor is connected across a voltage source of 120 V rms as shown in Fig. 2.43. A capacitor is connected across the load to improve the power factor to 0.95 lagging. Determine the capacitance of the connected capacitor.

**Solution**
The initial power factor is,

$$\cos \phi_1 = 0.75 \tag{2.271}$$

**Fig. 2.43**  Circuit for Example 2.18

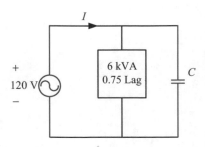

$$\phi_1 = 41.41° \tag{2.272}$$

The final power factor is,

$$\cos \phi_2 = 0.95 \tag{2.273}$$

$$\phi_2 = 18.19° \tag{2.274}$$

The real power is calculated as,

$$P = 6 \times 0.75 = 4.5 \, \text{kW} \tag{2.275}$$

The value of the parallel capacitance can be calculated as,

$$C = \frac{4.5 \times 1000(\tan 41.41° - \tan 18.19°)}{2\pi \times 50 \times 120^2} = 0.55 \, \text{mF} \tag{2.276}$$

**Example 2.19** A load of 10 kVA, 50 Hz, 0.8 lagging power factor is connected across the voltage source shown in Fig. 2.44. A capacitor is connected across the load to improve the power factor to 0.90 lagging. Calculate the value of the capacitance and line loss with and without the capacitor.

**Solution**
The initial power factor is,

$$\cos \phi_1 = 0.8 \tag{2.277}$$

$$\phi_i = 36.87° \tag{2.278}$$

The final power factor is,

**Fig. 2.44** Circuit for
Example 2.19

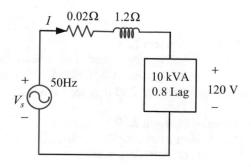

$$\cos \phi_2 = 0.9 \tag{2.279}$$

$$\phi_2 = 25.84° \tag{2.280}$$

The power of the load is calculated as,

$$P = 10 \times 0.8 = 8\,kW \tag{2.281}$$

The capacitor is calculated as,

$$C = \frac{8 \times 1000(\tan 36.87° - \tan 25.84°)}{2\pi \times 50 \times 120^2} = 0.47\,mF \tag{2.282}$$

The line current before adding capacitor is calculated as,

$$I_1 = \frac{8000}{0.8 \times 120} = 83.33\,A \tag{2.283}$$

The power loss in the line before adding capacitor is calculated as,

$$P_1 = 83.33^2 \times 0.02 = 138.88\,W \tag{2.284}$$

The apparent power with a power factor of 0.9 lagging is calculated as,

$$S = \frac{8000}{0.9} = 8888.89\,VA \tag{2.285}$$

The line current after adding capacitor is calculated as,

$$I_2 = \frac{8888.89}{120} = 74.07\,A \tag{2.286}$$

The power loss in the line after adding capacitor is calculated as,

$$P_2 = 74.07^2 \times 0.02 = 109.73\,W \tag{2.287}$$

**Practice Problem 2.18**
A load of 0.85 lagging power factor is connected across a voltage source of 220 V rms as shown in Fig. 2.45. A 0.56 mF capacitor is connected across the load to improve the power factor to 0.95 lagging. Find the value of the load, $P$.

**Practice Problem 2.19**
Two loads are connected to a source through a line as shown in Fig. 2.46. Determine the value of the voltage source.

**Fig. 2.45** Circuit for
practice problem 2.18

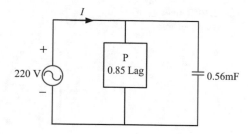

**Fig. 2.46** Circuit for
practice problem 2.19

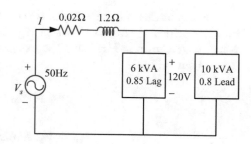

## 2.21 Three-Phase Voltage Generation

Figure 2.47 shows a two-pole three-phase AC generator for three-phase voltage
generation. Coils $aa'$, $bb'$ and $cc'$ represent the whole coils into a three-phase system
as shown in Fig. 2.47a. The rotor of the AC machine is energized by the dc source,
which creates the magnetic field. This rotor is attached to the turbine through a soft
coupling, and this turbine rotates the rotor. According to Faraday's law of electromagnetic induction, three-phase voltages $V_{an}$, $V_{bn}$ and $V_{cn}$ will be generated across
the generator terminals.

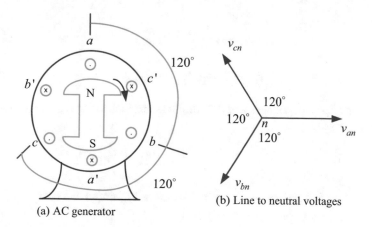

(a) AC generator

(b) Line to neutral voltages

**Fig. 2.47** Schematic of ac generator and phase voltages

**Fig. 2.48** Three-phase
voltage waveforms

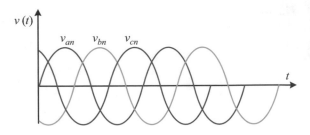

The magnitudes of these voltages are constant and are displaced from each other by 120 electrical degrees as shown in Fig. 2.47b. The waveforms of the generated voltages are shown in Fig. 2.48. The expression of the generated voltages can be represented as [1, 2],

$$v_{an} = V_{an} \sin \omega t \qquad (2.288)$$

$$v_{bn} = V_{bn} \sin(\omega t - 120°) \qquad (2.289)$$

$$v_{cn} = V_{cn} \sin(\omega t - 240°) \qquad (2.290)$$

where $V_{an}$, $V_{bn}$ and $V_{cn}$ are the magnitudes of the line to neutral or phase voltages. These voltages are constant in magnitude, and it can be expressed as,

$$|V_{an}| = |V_{bn}| = |V_{cn}| = V_p \qquad (2.291)$$

The phasor of the generated voltages can be written as,

$$V_{an} = V_P \underline{|0°} \qquad (2.292)$$

$$V_{bn} = V_P \underline{|-120°} \qquad (2.293)$$

$$V_{cn} = V_P \underline{|-240°} = V_P \underline{|120°} \qquad (2.294)$$

The sum of phasor voltages and the sum of sinusoidal voltages are zero, and these can be expressed as,

$$V = V_{an} + V_{bn} + V_{cn} = V_P \underline{|0°} + V_P \underline{|-120°} + V_P \underline{|-240°} = 0 \qquad (2.295)$$

$$v = v_{an} + v_{bn} + v_{cn} = V_{an} \sin \omega t + V_{bn} \sin(\omega t - 120°)$$
$$+ V_{cn} \sin(\omega t - 240°) = 0 \qquad (2.296)$$

**Fig. 2.49** Phase sequence
identification

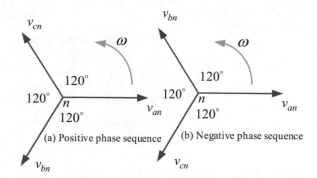

(a) Positive phase sequence    (b) Negative phase sequence

## 2.22   Phase Sequence

The phase sequence is very important for the interconnection of the three-phase
transformer, motor and other high voltage equipment. The three-phase systems are
numbered either by the numbers 1, 2 and 3 or by the letters *a, b,* and *c*. Sometimes,
these are labelled by the colours red, yellow and blue or *RYB* in short. The generator
is said to have a positive phase sequence when the generated voltages reach their
maximum or peak values in the sequential order of *abc*. Whereas the generator is said
to have a negative phase sequence when the generated voltages reach their maximum
or peak values in the sequential order of *acb*. Figure 2.49 shows the positive and
negative phase sequences.

Here, the voltage $V_{an}$ is considered to be the reference voltage while the direction
of rotation is considered to be anticlockwise. In the positive phase sequence, the
crossing sequence of voltage rotation is identified by $V_{an} - V_{bn} - V_{cn}$, whereas for
the negative phase sequence, it is identified as $V_{an} - V_{cn} - V_{bn}$.

## 2.23   Wye Connection

A three-phase transformer, AC generator and induction motor are connected either
in wye or in delta connection. In a wye connection, one terminal of each coil is
connected to form a common or neutral point, and other terminals are to the three-
phase supply. The voltage between any line and neutral is known as the phase voltage,
and the voltage between any two lines is called the line voltage. The line voltage and
phase voltage are usually represented by $V_L$ and $V_P$, respectively. The important
points of this connection are the line voltage is equal to $\sqrt{3}$ times the phase voltage,
the line current is equal to the phase current, and the current $(I_n)$ in the neutral wire
is equal to the phasor sum of the three-line currents. For a balanced three-phase load,
the neutral current is zero i. e., $I_n = 0$. The wye-connected generator and load are
shown in Fig. 2.50.

**Fig. 2.50** Wye connected
generator and load

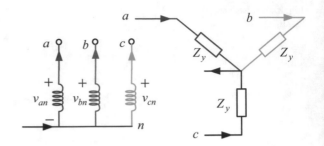

## 2.24   Analysis for Wye Connection

A three-phase wye-connected AC generator shown in Fig. 2.51 is considered for
analysis. Here, $V_{an}$, $V_{bn}$ and $V_{cn}$ are the phase voltages. Whereas $V_{ab}$, $V_{bc}$ and $V_{ca}$
are the line voltages. Applying KVL to the circuit to find the line voltage between
lines $a$ and $b$ yields,

$$V_{an} - V_{bn} - V_{ab} = 0 \tag{2.297}$$

$$V_{ab} = V_{an} - V_{bn} \tag{2.298}$$

Substituting Eqs. (2.292) and (2.293) into Eq. (2.298) yields,

$$V_{ab} = V_P \underline{/0°} - V_P \underline{/-120°} \tag{2.299}$$

$$V_{ab} = \sqrt{3}\, V_P \underline{/30°} \tag{2.300}$$

Applying KVL between lines $b$ and $c$ yields the expression of line voltage as,

**Fig. 2.51** Wye connected
generator

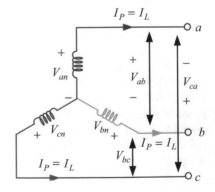

**Fig. 2.52** Phasor diagram
with line and phase voltages

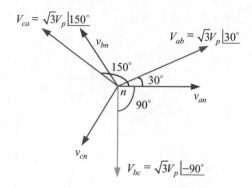

$$V_{bc} = V_{bn} - V_{cn} \tag{2.301}$$

Substituting Eqs. (2.293) and (2.294) into Eq. (2.301) yields,

$$V_{bc} = V_P \lfloor -120° - V_P \lfloor -240° = \sqrt{3}\, V_P \lfloor -90° \tag{2.302}$$

Applying KVL between lines $c$ and $a$ yield the expression of the line voltage as,

$$V_{ca} = V_{cn} - V_{an} \tag{2.303}$$

Substituting Eqs. (2.292) and (2.294) into Eq. (2.303) yields,

$$V_{ca} = V_P \lfloor -240° - V_P \lfloor 0° = \sqrt{3}\, V_P \lfloor 150° \tag{2.304}$$

Line voltages with angles are drawn as shown in Fig. 2.52. From Eqs. (2.300), (2.302) and (2.304), it is seen that the magnitude of the line voltage is equal to $\sqrt{3}$ times the magnitude of the phase voltage. The general relationship between the line voltage and the phase voltage can be written as,

$$V_L = \sqrt{3}\, V_P \tag{2.305}$$

From Fig. 2.51, it is also observed that the phase current is equal to the line current and it is written as,

$$I_L = I_P \tag{2.306}$$

**Alternative approach**: A vector diagram with phase voltages is drawn using the lines $a$ and $c$ as shown in Fig. 2.53. A perpendicular line is drawn from point $A$, which divides the line $BD$ equally. From the triangle $ABC$, the following expression relation can be written,

$$\cos 30° = \frac{BC}{AB} \tag{2.307}$$

**Fig. 2.53** A vector diagram

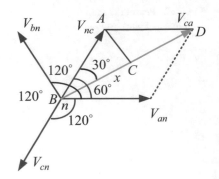

$$\frac{\sqrt{3}}{2} = \frac{x}{|Vnc|} \tag{2.308}$$

According to Fig. 2.53, the following expression can be written,

$$BD = 2AC \tag{2.309}$$

$$V_{ca} = 2x \tag{2.310}$$

Substituting Eq. (2.308) into Eq. (2.310) yields,

$$V_{ca} = 2 \times \frac{\sqrt{3}}{2}|V_{nc}| \tag{2.311}$$

$$V_{ca} = \sqrt{3}V_{nc} \tag{2.312}$$

In general, the following equation can be written as,

$$V_L = \sqrt{3}V_p$$

**Example 2.20** The phase voltage is given by $V_{an} = 230\underline{|10°}$ V. For *abc* phase sequence, determine $V_{bn}$ and $V_{cn}$.

**Solution**
The phase voltage for line *a* is calculated as,

$$V_{an} = 230\underline{|10°} \text{ V} \tag{2.313}$$

The phase voltage for line *b* is calculated as,

$$V_{bn} = 230\underline{|10° - 120°} = 230\underline{|-110°}\text{V} \tag{2.314}$$

**Fig. 2.54**  Circuit for
Example 2.21

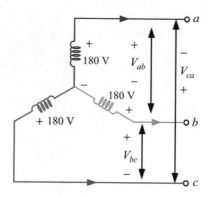

The phase voltage for line $c$ is calculated as,

$$V_{cn} = 230 \underline{|10° - 240°} = 230 \underline{|-230°} \text{ V} \tag{2.315}$$

**Practice Problem 2.20**
The phase voltage is given by $V_{bn} = 200 \underline{|10°}$ V. For $abc$ phase sequence, calculate
$V_{an}$ and $V_{cn}$.

**Example 2.21**  A wye-connected generator generates a voltage of 180 V rms as
shown in Fig. 2.54. For $abc$ phase sequence, write down the phase and line voltages.

**Solution**
The phase voltages are,

$$V_{an} = 180 \underline{|10°} \text{ V} \tag{2.316}$$

$$V_{bn} = 180 \underline{|-120°} \text{ V} \tag{2.317}$$

$$V_{cn} = 180 \underline{|-240°} \text{ V} \tag{2.318}$$

The line voltages are calculated as,

$$V_{ab} = \sqrt{3} \times 180 \underline{|30°} = 311.77 \underline{|30°} \text{ V} \tag{2.319}$$

$$V_{bc} = \sqrt{3} \times 180 \underline{|-120° + 30°} = 311.77 \underline{|-90°} \text{ V} \tag{2.320}$$

$$V_{ca} = \sqrt{3} \times 180 \underline{|-240° + 30°} = 311.77 \underline{|-210°} \text{ V} \tag{2.321}$$

**Fig. 2.55** Circuit for
practice problem 2.20

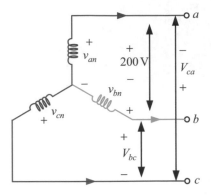

## Practice Problem 2.20

A wye-connected generator generates the line-to-line voltage of 200 V rms as shown
in Fig. 2.55. For *abc* phase sequence, write down the phase voltages.

## 2.25 Delta Connection

The coils in a delta-connected circuit are arranged in such a way that a looking
structure is formed. The delta-connection is formed by the connecting point $a_2$ of
$a_1a_2$ coil to the point $b_1$ of $b_1b_2$ coil, the point $b_2$ of $b_1b_2$ coil to the point $c_1$ of $c_1c_2$
coil and point $c_2$ of $c_1c_2$ coil to the point $a_1$ of $a_1a_2$ coil. In this connection, the phase
voltage is equal to the line voltage, and the line current is equal to $\sqrt{3}$ times the phase
current. Figure 2.56 shows delta-connected generator and load.

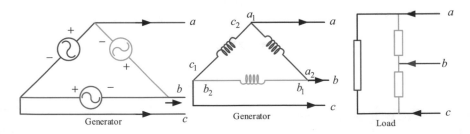

**Fig. 2.56** Delta connected generator and load

## 2.26   Analysis for Delta Connection

A three-phase delta-connected load is shown in Fig. 2.57. In this connection, $I_{ab}$, $I_{bc}$ and $I_{ca}$ are the phase currents and $I_a$, $I_b$ and $I_c$ are the line currents. For $abc$ phase sequence, the phase currents can be written as,

$$I_{ab} = I_P \underline{/0^\circ} \tag{2.322}$$

$$I_{bc} = I_P \underline{/-120^\circ} \tag{2.323}$$

$$I_{ca} = I_P \underline{/-240^\circ} \tag{2.324}$$

Applying KCL at the node $a$ of the circuit in Fig. 2.57 yields,

$$I_a = I_{ab} - I_{ca} \tag{2.325}$$

Substituting Eqs. (2.322) and (2.323) into Eq. (2.325) yields,

$$I_a = I_P \underline{/0^\circ} - I_P \underline{/-120^\circ} = \sqrt{3}\, I_P \underline{/30^\circ} \tag{2.326}$$

Applying KCL at the node $b$ of the circuit in Fig. 2.57 yields,

$$I_b = I_{bc} - I_{ab} \tag{2.327}$$

Substituting Eqs. (2.323) and (2.324) into Eq. (2.327) yields,

$$I_b = I_P \underline{/-120^\circ} - I_P \underline{/0^\circ} = \sqrt{3}\, I_P \underline{/-150^\circ} \tag{2.328}$$

Applying KCL at the node $c$ of the circuit in Fig. 2.57 yields,

$$I_c = I_{ca} - I_{bc} \tag{2.329}$$

Substituting Eqs. (2.322) and (2.324) into Eq. (2.329) yields,

**Fig. 2.57**  Delta connected load

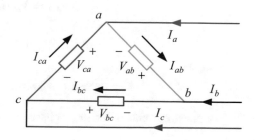

**Fig. 2.58** Phasor diagram
using line and phase currents

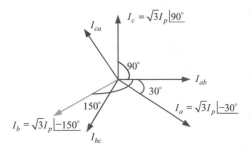

$$I_c = I_P \underline{\smash{-240°}} - I_P \underline{\smash{-120°}} = \sqrt{3}\ I_P \underline{\smash{-120°}} \underline{\smash{90°}} \tag{2.330}$$

From Eqs. (2.326), (2.328) and (2.330), it is found that the magnitude of the line current is equal to $\sqrt{3}$ times the phase current. The general relationship between the line current and the phase current is,

$$I_L = \sqrt{3}\ I_P \tag{2.331}$$

According to Fig. 2.57, it is observed that the phase voltage is equal to the line voltage i. e.,

$$V_L = V_P \tag{2.332}$$

The line and phase currents with their phase angles are drawn as shown in Fig. 2.58, where the phase current $I_{ab}$ is arbitrarily chosen as reference.

## 2.27   Analysis for Three-Phase Power

Consider a balanced three-phase wye-connected generator that delivers power to the balanced three-phase wye-connected load as shown in Fig. 2.59. The total power of the three-phase system is calculated by considering the instantaneous voltages and currents. The instantaneous voltages are,

$$v_{AN} = V_m \sin \omega t \tag{2.333}$$

$$v_{BN} = V_m \sin(\omega t - 120°) \tag{2.334}$$

$$v_{CN} = V_m \sin(\omega t - 240°) \tag{2.335}$$

The phase currents of the three-phase wye-connected load can be expressed as,

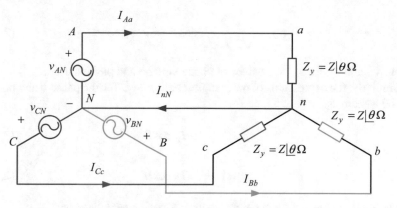

**Fig. 2.59** Wye-wye system for power calculation

$$i_{Aa} = \frac{v_{AN}}{Z_y} = \frac{V_m \ \sin \omega t}{Z \lfloor 0^\circ} = I_m \sin(\omega t - \theta) \tag{2.336}$$

$$i_{Bb} = \frac{v_{BN}}{Z_y} = \frac{V_m \ \sin(\omega t - 120^\circ)}{Z \lfloor \theta} = I_m \sin(\omega t - \theta - 120^\circ) \tag{2.337}$$

$$i_{Cc} = \frac{v_{CN}}{Z_y} = \frac{V_m \ \sin(\omega t - 240^\circ)}{Z \lfloor \theta} = I_m \sin(\omega t - \theta - 240^\circ) \tag{2.338}$$

The instantaneous power for phase $a$ can be expressed as [8, 9],

$$p_a(t) = \frac{1}{T} \int_0^T v_{AN} i_{Aa} \, dt \tag{2.339}$$

Substituting Eqs. (2.333) and (2.336) into Eq. (2.339) yields,

$$p_a(t) = \frac{V_m I_m}{T} \int_0^T \sin \omega t \times \sin(\omega t - \theta) \, dt \tag{2.340}$$

$$p_a(t) = \frac{V_m I_m}{2T} \int_0^T 2 \sin \omega t \times \sin(\omega t - \theta) \, dt \tag{2.341}$$

$$p_a(t) = \frac{V_m I_m}{2T} \int_0^T [\cos \theta - \cos(2\omega t - \theta)] \, dt \tag{2.342}$$

$$p_a(t) = \frac{V_m I_m}{2T} \times \cos \theta \times T - 0 \tag{2.343}$$

$$p_a(t) = \frac{V_m I_m}{\sqrt{2} \times \sqrt{2}} \cos\theta = V_P I_P \cos\theta \qquad (2.344)$$

where $V_p$ and $I_p$ are the rms values of phase voltage and phase current.

Similarly, the expressions of the instantaneous power for the phase $b$ and phase $c$ can be written as,

$$p_b(t) = V_P I_P \cos\theta \qquad (2.345)$$

$$p_c(t) = V_P I_P \cos\theta \qquad (2.346)$$

Therefore, the average three-phase power $P$ can be calculated as,

$$P = p_a(t) + p_b(t) + p_c(t) \qquad (2.347)$$

Substituting Eqs. (2.344), (2.345) and (2.346) into Eq. (2.347) yields,

$$P_t = 3V_p I_p \cos\theta \qquad (2.348)$$

Similarly, the expression of three-phase reactive power can be expressed as,

$$Q_t = 3\,V_p I_p \sin\theta \qquad (2.349)$$

Therefore, the per phase average ($P_{pp}$) and reactive ($Q_{pp}$) power can be written as,

$$P_{pp} = V_P I_P \cos\theta \qquad (2.350)$$

$$Q_{pp} = V_P I_P \sin\theta \qquad (2.351)$$

The complex power per phase $S_{pp}$ is represented as,

$$S_{pp} = P_{pp} + j Q_{pp} \qquad (2.352)$$

Substituting Eqs. (2.350) and (2.351) into Eq. (2.352) yields,

$$S_{pp} = V_P I_P \cos\theta + j V_P I_P \sin\theta \qquad (2.353)$$

Equation (2.353) can be expressed as,

$$S_{pp} = V_P I_P \underline{|\theta} \qquad (2.354)$$

From Eq. (1.354), it is seen that the per phase complex power is equal to the product of the voltage per phase and the phase current with an angle.

**Y-connection:** Substituting Eqs. (2.305) and (2.306) into Eq. (2.348) yields,

$$P_{tY} = 3 \times \frac{V_L}{\sqrt{3}} \times I_L \times \cos\theta \qquad (2.355)$$

$$P_{tY} = \sqrt{3}\,V_L I_L \cos\theta \qquad (2.356)$$

Substituting Eqs. (2.305) and (2.306) into Eq. (2.349) yields,

$$Q_{tY} = 3 \times \frac{V_L}{\sqrt{3}} \times I_L \times \sin\theta \qquad (2.357)$$

$$Q_{tY} = \sqrt{3}\,V_L I_L \sin\theta \qquad (2.358)$$

**Delta connection**: Again, substituting Eqs. (2.331) and (2.332) into Eq. (2.348) yields,

$$P_{t\Delta} = 3 \times \frac{I_L}{\sqrt{3}} \times V_L \times \cos\theta \qquad (2.359)$$

$$P_{t\Delta} = \sqrt{3}\,V_L I_L \cos\theta \qquad (2.360)$$

Substituting Eqs. (2.331) and (2.332) into Eq. (2.349) yields,

$$Q_{t\Delta} = 3 \times \frac{V_L}{\sqrt{3}} \times I_L \times \sin\theta \qquad (2.361)$$

$$Q_{t\Delta} = \sqrt{3}\,V_L I_L \sin\theta \qquad (2.362)$$

In general, the total real and reactive power can be expressed as,

$$P_t = \sqrt{3}\,V_L I_L \cos\theta \qquad (2.363)$$

$$Q_t = \sqrt{3}\,V_L I_L \sin\theta \qquad (2.364)$$

The total complex power can be written as,

$$S_t = P_t + jQ_t \qquad (2.365)$$

Substituting Eqs. (2.363) and (2.364) into Eq. (2.365) yields,

$$S_t = \sqrt{3}\,V_L I_L \cos\theta + j\sqrt{3}\,V_L I_L \sin\theta \qquad (2.366)$$

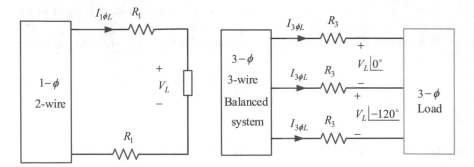

**Fig. 2.60** Single-phase and three-phase systems with loads

$$S_t = \sqrt{3} \, V_L I_L \underline{|\theta}$$  (2.367)

A three-phase system uses less amount of copper wire than the single-phase system for the same line voltage and same power factor to transmit the same amount of power over a fixed distance. From Fig. 2.60, the real power for a single-phase two-wire system is,

$$P_{1\phi 2w} = V_L I_{1\phi L} \cos \theta$$  (2.368)

From Fig. 2.60, the real power for a three-phase three wire system is,

$$P_{3\phi 3w} = \sqrt{3} \, V_L I_{3\phi L} \cos \theta$$  (2.369)

Equations (2.368) and (2.369) will be equal for transmitting or delivering same amount of power over a fixed distance. It can be expressed as,

$$V_L I_{1\phi L} \cos \theta = \sqrt{3} \, V_L I_{3\phi L} \cos \theta$$  (2.370)

$$I_{1\phi L} = \sqrt{3} \, I_{3\phi L}$$  (2.371)

The power loss in the single-phase wire is,

$$P_{1\phi 2wloss} = 2 I_{1\phi L}^2 R_1$$  (2.372)

$$P_{3\phi 3wloss} = 3 I_{3\phi L}^2 R_3$$  (2.373)

From Eqs. (2.372) and (2.373), the ratio of power loss of a single-phase system to a three-phase system can be derived as,

$$\frac{P_{1\phi2wloss}}{P_{3\phi3wloss}} = \frac{2I_{1\phi L}^2 R_1}{3I_{3\phi L}^2 R_3} \tag{2.374}$$

For equal losses ($P_{1\phi2wloss} = P_{3\phi3wloss}$), Eq. (2.374) can be modified as,

$$1 = \frac{2I_{1\phi L}^2 R_1}{3I_{3\phi L}^2 R_3} \tag{2.375}$$

$$\frac{3I_{3\phi L}^2}{2I_{1\phi L}^2} = \frac{R_1}{R_3} \tag{2.376}$$

Substituting Eq. (2.373) into Eq. (2.376) yields,

$$\frac{3I_{3\phi L}^2}{2 \times 3I_{3\phi L}^2} = \frac{R_1}{R_3} \tag{2.377}$$

$$\frac{R_1}{R_3} = \frac{1}{2} \tag{2.378}$$

The following ratio can be written as,

$$\frac{\text{Copper for } 3\phi \text{ system}}{\text{Copper for } 1\phi \text{ system}} = \frac{\text{number of wires in } 3\phi \text{ system}}{\text{number of wires in } 1\phi \text{ system}} \times \frac{R_1}{R_3} \tag{2.379}$$

Substituting Eq. (2.378) and the number of wires for both systems in Eq. (2.379) yields,

$$\frac{\text{Copper for } 3\phi \text{ system}}{\text{Copper for } 1\phi \text{ system}} = \frac{3}{2} \times \frac{1}{2} \tag{2.380}$$

$$\text{Copper for } 3\phi \text{ system} = \frac{3}{4} \times \text{Copper for } 1\phi \text{ system} \tag{2.381}$$

From Eq. (2.381), it is seen that the copper required for a three-phase system is equal to the three-fourths of the copper required for a single-phase system.

**Example 2.21** A balanced three-phase wye-wye system is shown in Fig. 2.61. For ABC phase sequence, calculate the line current, power supplied to each phase, power absorbed by each phase and the total complex power supplied by the source.

**Solution**
The line currents are calculated as,

$$I_{Aa} = \frac{230\underline{|15°}}{2 + j8} = 27.89\underline{|-60.96°} \text{ A} \tag{2.382}$$

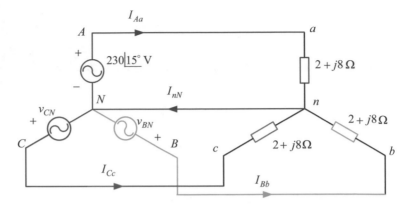

**Fig. 2.61**  Circuit for Example 2.21

$$I_{Bb} = I_{Aa} \underline{|-120°} = 27.89 \underline{|-180.96°} \text{ A} \qquad (2.383)$$

$$I_{Cc} = I_{Aa} \underline{|+120°} = 27.89 \underline{|59.04°} \text{ A} \qquad (2.384)$$

Power supplied to each phase is calculated as,

$$
\begin{aligned}
P_A &= V_p I_p \cos\theta = V_{AN} I_{Aa} \cos(\theta_v - \theta_i) = 230 \times 27.89 \\
&\quad \times \cos(15 + 60.96) = 1556.20 \text{ W}
\end{aligned} \qquad (2.385)
$$

Per phase power absorbed by the load is calculated as,

$$P_{L1\phi} = 27.89^2 \times 2 = 1555.70 \text{ W} \qquad (2.386)$$

The total complex power supplied by the source is calculated as,

$$S_t = 3V_{An} I_{Aa}^* = 3 \times 230 \underline{|-15°} \times 27.89 \underline{|60.96°} = 4668.60 + j18669.21 \text{ VA} \qquad (2.386)$$

**Practice Problem 2.21**

A balanced three-phase wye-delta system is shown in Fig. 2.62. For *ABC* phase sequence, calculate the line current, power supplied to each phase, power absorbed by each phase and the total complex power supplied by the source.

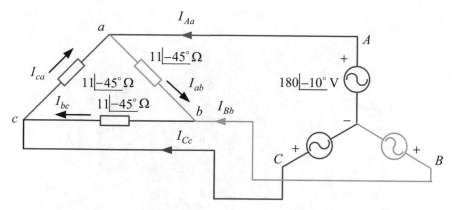

**Fig. 2.62** Circuit for practice problem 2.21

## 2.28 Basic Measuring Equipment

In laboratory experiments, students usually verify fundamental electrical theories through different types of measurement of voltage, current, resistance and power. The voltmeter, ammeter, ohmmeter and wattmeter are used to measure those parameters. Nowadays, almost every educational laboratory is equipped with digital meters. An ohmmeter is used to check the circuit continuity, identify short circuits and open circuits, find out the specific lead of a multi-lead cable. The voltmeter is used to measure the voltage of a circuit, and it always connects in parallel with a specific element. The voltmeter has high input impedance. An ammeter is an instrument that is used to measure the current in the circuit. An ammeter has low input impedance and connects in series with an element of the circuit. Under the energized condition, the ammeter cannot disconnect from the circuit, whereas the voltmeter can be disconnected from the circuit. A wattmeter has two coils, namely a voltage coil and a current coil. The voltage coil is connected across the element, and the current coil is connected in series. The symbols of voltmeter, ammeter, wattmeter and ohmmeter are shown in Fig. 2.63.

There are many advanced electrical meters available for use in the practical field and can be obtained at a low price from hardware stores. However, companies such as Fluke Inc. manufactured digital multimeters as shown in Fig. 2.64, which includes

**Fig. 2.63** Symbols of basic electrical meters

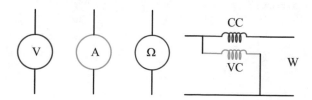

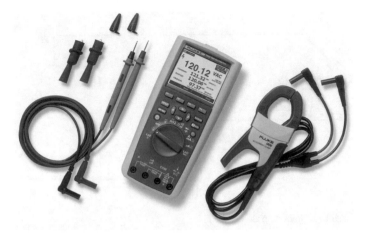

**Fig. 2.64** Electrical meters courtesy by FLUKE Inc

advanced safety features during practical measurement. These meters have propri-
etary functions that prevent any accidents resulting from breakers that suddenly
trip due to incorrect connections. These meters also have high accuracy, reliability,
extensive additional functionality and a broad range of measurement activities.

**Exercise Problems**

2.1   The excitation voltage and the impedance of a series circuit are given by
      $v(t) = 8 \sin 10t$ V and $Z = 5 \underline{|10°}$ $\Omega$, respectively. Calculate the instantaneous
      power.

2.2   The excitation current and the impedance of a series circuit are given by
      $i(t) = 4 \sin(100t - 20°)$A and $Z = 5 \underline{|10°}$ $\Omega$, respectively. Determine the
      instantaneous power.

2.3   Calculate the average power supplied by the source and the power absorbed
      by the resistors as shown in Fig. P2.1.

2.4   Determine the average power supplied by the source and the power absorbed
      by the 8$\Omega$ resistor shown in Fig. P2.2.

**Fig. P2.1** Circuit for
problem 1.3

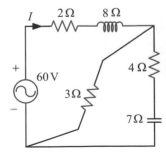

**Fig. P2.2** Circuit for problem 1.4

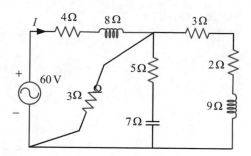

2.5 Calculate the average power supplied by the source and the power absorbed by the 3Ω resistor as shown in Fig. 2.3.

2.6 Determine the average power supplied by the source and the power absorbed by the 2Ω resistor shown in Fig. P2.4.

2.7 Find the total average power absorbed by all the resistors in the circuit shown in Fig. P2.5.

2.8 An industrial load is connected across an alternating voltage source $v(t) = 230 \sin(314t + 20°)$ V that draws a current of $i(t) = 15 \sin(314t + 45°)$ A. Determine the apparent power, circuit resistance and capacitance.

2.9 The rms values of voltage and current are given by $V = 20\lfloor-15°$ V, and $I = 3\lfloor25°$ A. Calculate the complex power, real power and reactive power.

**Fig. P2.3** Circuit for problem 2.5

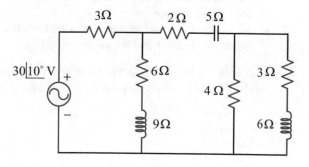

**Fig. P2.4** Circuit for problem 2.2

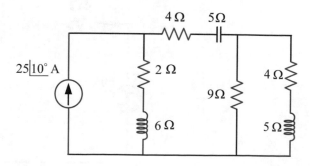

**Fig. P2.5** Circuit for
problem 2.7

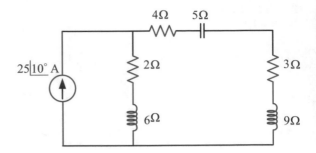

2.10   The rms values of voltage is given by $V = 34\underline{|25°}$ V, and the impedance
      is $Z = 6\underline{|-15°}$ Ω. Determine the complex power, real power and reactive
      power.

2.11   A series–parallel circuit is supplied by an rms source of 60 V as shown in
      Fig. P2.2. Find the total complex power.Fig. P2.6

2.12   Calculate the total complex power of the circuit shown in Fig. P2.7.

2.13   A 10 kVA, 50 Hz, 0.6 lagging power factor load is connected across an rms
      voltage source of 220 V as shown in Fig. P2.8. A capacitor is connected
      across the load to improve the power factor to 0.85 lagging. Find the value
      of the capacitor.

2.14   Two loads with different power factors are connected with the source through
      a transmission line as shown in Fig. P2.9. Determine the source current and
      the source voltage.

12.15  A voltage source delivers power to the three loads shown in Fig. P2.10. Find
      the source current and the source voltage.

2.16   The line voltage of a three-phase wye-connected generator is found to be
      440 V. For *abc* phase sequence, calculate the phase voltages.

**Fig. P2.6** Circuit for
problem 2.11

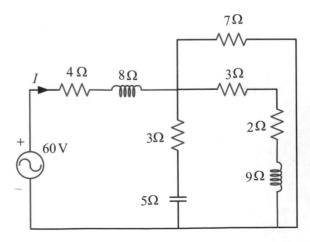

**Fig. P2.7** Circuit for
problem 2.12

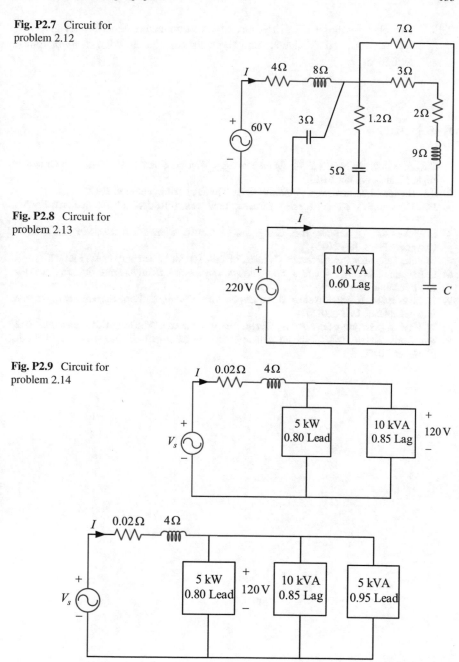

**Fig. P2.8** Circuit for
problem 2.13

**Fig. P2.9** Circuit for
problem 2.14

**Fig. P2.10** Circuit for problem 2.15

2.17   The phase voltage of a three-phase wye-connected generator is given by $V_{an} = 100\underline{/-10°}$ V. Determine the voltages $V_{bn}$ and $V_{cn}$ for $abc$ phase sequence.

# References

1. C.K. Alexander, M.N.O. Sadiku, *Fundamentals of Electric Circuits*, Sixth edn. (McGraw-Hill Higher Education, Jan 2012)
2. R.L. Boylestad, *Introductory Circuit Analysis*, Thirteenth edn. (Pearson, 2012)
3. J.W. Nilsson, S.A. Riedel, *Electric Circuits*, Tenth edn. (Prentice-Hall International Edition, 2015)
4. H.W. Jackson, D. Temple, B.E. Kelly, *Introduction to Electric Circuits*, 9th edn. (Oxford University Press, July 2015)
5. D. Bell, *Fundamentals of Electric Circuits*, 7th edn. (Oxford University Press, Mar 2007)
6. G. Rizzoni, J. Kearns, *Principles and Applications of Electrical Engineering*, 6th edn (McGraw-Hill Education, Oct 2014)
7. J. David Irwin, R. Mark Nelms, *Basic Engineering Circuit Analysis*. Eleventh edn. (John & Wiley Sons Inc., USA, 2015)
8. W. Hayt, J. Kemmerly, *Engineering Circuit Analysis*, 8th edn. (McGraw-Hill Education, 2012)
9. Md. Abdus Salam, Q.M. Rahman, *Fundamentals of Electrical Circuits Analysis*, First edn. (Springer, 2018)

# Chapter 3
# Fluid Power Properties

## 3.1 Introduction

The properties of the fluid are important to operate the fluid power system. The fluid can be used as compressible and incompressible. Compressible fluids such as air and gas are used in the pneumatics circuit operation. Whereas, incompressible fluids such as water and petroleum are used in the operation of a hydraulic circuit. This chapter will discuss mass, density, specific weight and gravity, viscosity, pascal law, continuity equation, Bernoulli's equation, Darcy's equation, Colebrook equation.

## 3.2 Mass and Weight

Mass is the fundamental property of an object and simply how much matter stuff an object consists of and it is denoted by $m$. The common units of mass are kg, slugs. The mass of the object is the same anywhere in the universe. The mass is how something is to a given force. From Newton's second law, the mass is expressed as [1],

$$m = \frac{F}{a} \tag{3.1}$$

where,

$F$ is the force in units of Newton (N),
$a$ is the acceleration in units of m/s$^2$,
$m$ is the mass in units of kg.

All objects either solids or fluids are pulled toward the center of the earth by a force of attraction that is called the weight of the object. In other words, weight is the amount of force that an object's mass generates due to gravity. The weight, $W$, of a fluid, is defined as the mass multiplied by the gravitational acceleration, $g$.

© The Author(s), under exclusive license to Springer Nature Singapore Pte Ltd. 2022
Md. A. Salam, *Fundamentals of Pneumatics and Hydraulics*,
https://doi.org/10.1007/978-981-19-0855-2_3

Mathematically, it can be expressed as,

$$F = W = mg \tag{3.2}$$

where,

$F$ is the force in units of lb,
$W$ is the weight in units of lb,
$m$ is the mass of the object in units of slugs,
$g$ is the proportionality constant called the acceleration of gravity, which equals 33.2 ft/s$^2$ and 9.81 m/s$^2$ at sea level.

**Example 3.1**  The mass of an object is found to be 30 kg. Calculate its weight.

**Solution**
The value of the weight is calculated as,

$$W = mg = 30 \, (\text{kg}) \times 9.81 \, (\text{m/s}^2) = 294.3 \, \text{kg-m/s}^2 = 294.3 \, \text{N} \tag{3.3}$$

**Example 3.2**  The weight of an object is found to be 500 lb. Determine its mass.

**Solution**
The value of the mass of an object is determined as,

$$m = \frac{W}{g} = \frac{500}{32.2} = 15.58 \, \text{lb-ft/s}^2 = 15.58 \, \text{slugs} \tag{3.4}$$

**Practice Problem 3.1**
The mass of an object is found to be 150 kg. Find its weight.

**Practice Problem 3.2**
The weight of an object is found to be 1500 lb. Calculate its mass.

## 3.3  Density

Any object has an identical density regardless of its shape, size and mass. Let us consider that an object has $n$ molecules per atom as shown in Fig. 3.1. Then the density can be expressed as,

$$\rho = \frac{nm}{l \times l \times l} = \frac{M(m)}{V} \tag{3.5}$$

**Fig. 3.1** An object with a cubic shape

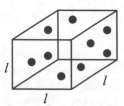

Therefore, the heaviness of an object refers to density. In other words, the mass per unit volume of a fluid is known as density. The density is denoted by the Greek letter $\rho$. Its units are $kg/m^3$, $slugs/in.^3$, $slugs/ft^3$. A typical density for a hydraulic fluid (oil) is 3.74 $slugs/ft^3$ or 897 $kg/m^3$.

A certain object is said to be denser than another if that object has more atoms in a given volume. Again, from Eq. (3.5), the mass can be expressed as,

$$m = \rho V \tag{3.6}$$

For example, oil and boat, both float in the water because it is less dense than water. The density of some parameters is shown in Table 3.1.

***Example 3.3*** The radius and density of a solid steel ball are 20 cm and 7.8 × 103 $kg/m^3$, respectively. Calculate its mass.

**Solution**
The value of the mass of a solid steel ball is determined as,

$$m = \rho V = \rho\left(\frac{4}{3}\pi r^3\right) = 7.8 \times 10^3 \times \left(\frac{4}{3} \times \pi \times 0.2^3\right) = 261.38\,kg \tag{3.7}$$

**Practice Problem 3.3**
A container is filled with a liquid whose weight is found to be 1500 pounds. The specification of a container is 3 ft long, 2 ft wide, and 4 ft deep. Find the density.

**Table 3.1** Density for a few parameters

| Parameter | Density ($kg/m^3$) |
| --- | --- |
| Water | 1000 |
| Dry sand | 1600 |
| Steel | 7849 |
| Mercury | 13,600 |
| Cement | 1440 |

## 3.4  Specific Weight

The specific weight in a fluid power system is usually related to the volume and
its weight. To explain the volume, consider a cubic container that contains the fluid
(water) as shown in Fig. 3.2. The length, base and height of the container are the
same and it is conserved as $l = 1$ ft.

In fluid mechanics, the specific weight represents the force exerted by gravity on
a unit volume of the fluid. The weight per unit volume is known as specific weight.
It is denoted by a Greek letter $\gamma$ and its units are N/m³, lb/ft³, etc. Mathematically,
the specific weight is expressed as [2],

$$\gamma = \frac{W}{V} \tag{3.8}$$

where,

$W$ is the weight in the units of N, lb,
$V$ is the volume in the unit of m³, ft³.
$\gamma$ is the specific weight in the units N/m³, lb/ft³.

The specific weight. on the other hand, is not absolute, since it depends on the
value of gravitational acceleration ($g$), which varies with location, primarily latitude
and elevation above mean sea level. The specific weight at 20 °C wth $g = 9.81$ m/s²
is shown in Table 3.2.

**Fig. 3.2**  A cubic shape
container

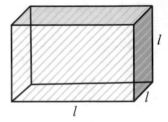

**Table 3.2**  Specific weight at
20 °C wth g = 9.81 m/s³

| Parameter | $\gamma$ (kN/m³) |
| --- | --- |
| Water | 9.79 |
| Seawater | 10.03 |
| Gasoline | 6.6 |
| Glycerin | 13.3 |
| Kerosene | 7.9 |
| Motor oil | 8.5 |

***Example 3.4*** The diameter and the height of a cylinder container are 0.4 m and 3.5 m, respectively. The container is filled with a 15 kg liquid. Calculate the specific weight.

**Solution**

The value of the volume is calculated as,

$$V = Ah = \pi \frac{d^2}{4} h = \pi \frac{0.4^2}{4} \times 1.5 = 0.188 \, \text{m}^3 \tag{3.9}$$

The value of the specific weight is calculated as,

$$\gamma = \frac{W}{V} = \frac{15}{0.188} = 79.79 \, \text{N/m}^3 \tag{3.10}$$

**Practice Problem 3.4**

A cylinder container has a diameter of 0.2 m and a height of 2 m. If it is filled with a liquid having a specific weight of 1500 N/m$^3$, how many kg of this liquid must be added to fill the container?

## 3.5  Specific Gravity

The specific gravity is very important in an automobile battery where sulphuric acid ($H_2SO_4$) is mixed with water to get an electrolyte solution. Therefore, the specific gravity is always with water as a reference. Specific gravity is defined as the ratio of the specific weight of an object to the specific weight of the water and it has no unit. The expression of specific weight is modified as,

$$\gamma = \frac{W}{V} = \frac{mg}{V} \tag{3.11}$$

Substituting Eq. (3.5) into Eq. (3.11) yields,

$$\gamma = \rho g \tag{3.12}$$

Mathematically, the specific gravity is expressed as,

$$S = \frac{\gamma_{\text{object}}}{\gamma_{\text{water}}} \tag{3.13}$$

Substituting Eq. (3.12) into Eq. (3.13) yields,

$$S = \frac{g\rho_{\text{object}}}{g\rho_{\text{water}}} = \frac{\rho_{\text{object}}}{\rho_{\text{water}}} \tag{3.14}$$

Therefore, the specific gravity for oil and air can be written as,

$$S_{oil} = \frac{\rho_{oil}}{\rho_{water}} \tag{3.15}$$

$$S_{air} = \frac{\rho_{air}}{\rho_{water}} \tag{3.16}$$

If the specific gravity of a liquid is greater than 1 ($S > 1$), it is thicker than water, it sinks in water. If the specific gravity of a liquid is less than 1 ($S < 1$), it is thinner than water, it floats on water.

***Example 3.5***  One litre ($1/1000\,\text{m}^3$) of SAE30 oil weighs 0.75 N. Calculate its specific weight, density and specific gravity.

**Solution**
The value of the specific weight is calculated as,

$$\gamma = \frac{W}{V} = \frac{0.75}{1/1000} = 750\,\text{N/m}^3 \tag{3.17}$$

The value of the density is calculated as,

$$\rho_{oil} = \frac{\gamma}{g} = \frac{750}{9.81} = 76.45\,\text{kg/m}^3 \tag{3.18}$$

The value of the specific gravity is determined as,

$$S_{oil} = \frac{\rho_{oil}}{\rho_{water}} = \frac{76.45}{1000} = 0.076 \tag{3.19}$$

**Practice Problem 3.5**
The density of gold is $19300\,\text{kg/m}^3$ and the density of water is $1000\,\text{kg/m}^3$. Calculate the specific gravity.

## 3.6   Pressure

The pressure and the force are the two important parameters that are normally used in the actuators and pumps operation. The pressure is defined as the force per unit area. In other words, pressure is the amount of force acting over a unit area. Mathematically, it can be expressed as,

$$p = \frac{F}{A} \tag{3.20}$$

**Fig. 3.3** Cubic container for
any liquid

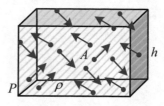

The liquid direction inside a cubic container is shown in Fig. 3.3. The pressure developed at the bottom of a container of any liquid is known as hydrostatic pressure and it is expressed as,

$$p = \rho g h \tag{3.21}$$

Substituting Eq. (3.5) into Eq. (3.21) yields,

$$p = \frac{mg}{V} h \tag{3.22}$$

Again, substituting Eq. (3.2) into Eq. (3.22) yields,

$$p = \frac{W}{V} h \tag{3.23}$$

Substituting Eq. (3.8) into Eq. (3.23) yields,

$$p = \gamma h \tag{3.24}$$

From Eq. (3.24), it is seen that the value of the pressure is can be calculated if the specific weight and the height are known.

## 3.7  Bottom Pressure of Liquid Column

The pressure will come from water and act on all sides of any object when floating into water. Therefore, when floating a boat into the water, the boat is subjected to pressure coming from all sides of the water. Consider a container with a liquid. The area and the volume of the liquid are $A$ and $V$, respectively. The pressure head developed at the bottom due to the column of liquid. Let $h$ be the height of the liquid column and $W$ be the weight of the liquid as shown in Fig. 3.4.

The liquid has a specific weight and volume. From the basic definition of pressure, the following relation can be expressed as,

$$p = \frac{F}{A} \times \frac{V}{V} = \frac{F}{V} \times \frac{V}{A} \tag{3.25}$$

**Fig. 3.4** A liquid column

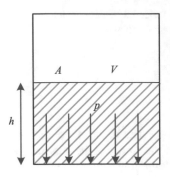

Substituting $F = W$ and $V = Ah$ into Eq. (3.25) yields,

$$p = \frac{W}{V} \times \frac{Ah}{A} \tag{3.26}$$

Substituting Eq. (3.8) into Eq. (3.26) yields,

$$p = \gamma h \tag{3.27}$$

**Example 3.6** A tank is filled with oil whose specific weight is 5000 N/m³. If the depth of the tank is 2 m, find the pressure at the bottom of the tank.

**Solution**
The value of the pressure at the bottom of the tank is calculated as,

$$P = \gamma h = 5000 \times 2 = 10,000 \, \text{N/m}^2 \, (10 \, \text{kPa}) \tag{3.28}$$

**Practice Problem 3.6**
The pressure at the bottom of an oil-filled tank is 15 kPa and the specific weight is 2500 N/m³. Determine the depth of the tank.

## 3.8   Viscosity

The fluid is used to transmit the power in a fluid power system. The fluid is also used to lubricant both the hydraulic and pneumatic systems. The word 'Viscosity' is derived from the Latin word 'VISCOUM' meaning 'MISTLETOE' and 'VISCOUS GLUE' made from mistletoe berries. Viscosity is an important fluid property when analyzing liquid behaviour and fluid motion near solid boundaries. Viscosity is known as the internal friction of a moving fluid.

Viscosity is a quantitative measure of a fluid's resistance to flow. In an electrical circuit, resistance is the analogy of viscosity in fluid power. High resistance reduces the flow of electric current and low resistance increases the flow of current in a circuit. Similarly, when the viscosity is low, the fluid can flow easily, and it is thin in appearance. Whereas, when the viscosity is high, the fluid flow is reduced, and it is thick in appearance. High viscosity increases the power losses due to frictional losses and sluggish operation. It also increases pressure drop through the vales and lines.

Viscosity is classified as dynamic or absolute viscosity and kinematic viscosity.

### 3.8.1 Dynamic Viscosity

The dynamic or absolute viscosity of a fluid is a measure of internal resistance it offers to relative shearing motion. In other words, the ratio of shearing stress in oil and the slope of the velocity profile is known as dynamic or absolute viscosity. It is represented by a Greek letter $\mu$ and its unit is N-s/m$^2$ or lb-s/ft$^{3.}$ Mathematically, it can be expressed as,

$$\mu = \frac{\text{Shear stress in oil}}{\text{Slope of velocity profile}} = \frac{\tau}{\frac{v}{y}} \qquad (3.29)$$

Consider that the top plate is moving at a velocity $v$ as it push by a force $F$ and the bottom plate is stationary as shown in Fig. 3.5.

The shearing stress is defined as the force per unit area. Now, substituting $\tau = F/A$ in Eq. (3.29) yields,

$$\mu = \frac{\frac{F}{A}}{\frac{v}{y}} \qquad (3.30)$$

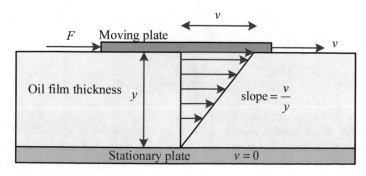

**Fig. 3.5** Two plates with oil film thickness

$$\mu = \frac{F \times y}{A \times v} \tag{3.31}$$

In CGS (centimeter, gram, second) system, the most common unit for the dynamic viscosity is centipoise (cP), which is equivalent to 0.01 Poise (P) in honor of French physicist, Jean Leonard Marie Poiseuille (1797–1869).

### 3.8.2   Kinematic Viscosity

The kinematic viscosity is used to identify the relationship between the viscous force and the internal force in a fluid. The ratio of absolute viscosity to the density of the fluid is known as kinematic viscosity and it is denoted by a Greek letter $v$ (nu). In a hydraulic system, kinematic viscosity is used for calculations rather than an absolute viscosity. The SI unit of kinematic viscosity is $m^2/s$. In the CGS system, the kinematic viscosity is often measured in $cm^2/s$, after Irish mathematician and inventor of viscosity Sir George Gabriel Stokes (1819–1903). The unit of viscosity $cm^2/s$ is called the Stoke (S), and many scientists used centistoke (cS), which is equivalent to 0.01 Stoke (S). Mathematically, the kinematic viscosity is expressed as,

$$v(nu) = \frac{\mu}{\rho} \tag{3.32}$$

Again, the kinematic viscosity in cS is the ratio of absolute viscosity in cP to the specific gravity and it is expressed as,

$$v(cS) = \frac{\mu(cP)}{\rho} \tag{3.33}$$

**Example 3.7**  The absolute and kinematic viscosities of fluid are given as $0.85 \, Ns/m^2$ and $5 \, m^2/s$. Calculate the density of a fluid.

**Solution**
The value of the density of the fluid is calculated as,

$$\rho = \frac{\mu}{v} = \frac{0.85}{5} = 0.17 \, kg/m^3 \tag{3.34}$$

**Example 3.8**  A 3.5 N force moves a piston inside a cylinder at a velocity of 1.5 m/s. The diameters of the cylinder and piston are 9.5 cm and 9.4 cm, respectively. An oil film separates the piston from the cylinder and the length of the piston is 4 cm. Calculate the absolute viscosity of the oil.

**Solution**

The value of the area of the piston is calculated as,

$$A = \pi DL = \pi \times 9.4 \times 4 = 590.61\,\text{cm}^2 = \frac{590.61}{100 \times 100} = 0.059\,\text{m}^2 \qquad (3.35)$$

The value of the oil film thickness is calculated as,

$$y = \frac{D - d}{2} = \frac{9.5 - 9.4}{2} = 0.25\,\text{cm} = \frac{0.25}{100} = 2.5 \times 10^{-3}\,\text{m} \qquad (3.36)$$

The value of the dynamic viscosity is calculated as,

$$\mu = \frac{F \times y}{A \times v} = \frac{3.5 \times 2.5 \times 10^{-3}}{0.059 \times 1.5} = 0.098\,\text{Ns/m}^2 \qquad (3.37)$$

**Practice Problem 3.7**

The absolute viscosity and density of fluid are given as 0.95 Ns/m² and 3.5 kg/m³. Calculate the kinematic viscosities of a fluid.

**Practice Problem 3.8**

A 5 N force moves a piston inside a cylinder at a velocity of 3.5 m/s. The diameters of the cylinder and piston are 8 cm and 7.9 cm, respectively. An oil film separates the piston from the cylinder and the length of the piston is 5.2 cm. Calculate the absolute viscosity of the oil.

## 3.9 Viscosity Index

The viscosity of hydraulic oil is inversely related to the temperature. The viscosity of hydraulic oils decreases with an increase in temperature as shown in Fig. 3.6. Therefore, the variation of viscosity with respect to temperature is different for different types of oil.

In general, the viscosity index (VI) is a relative measure of the change in the viscosity of oil with respect to a change in temperature. The VI of any hydraulic oil can be expressed as,

$$VI = \frac{L - U}{L - H} \times 100 \qquad (3.38)$$

where,

**Fig. 3.6** Variation of
viscosity with temperature

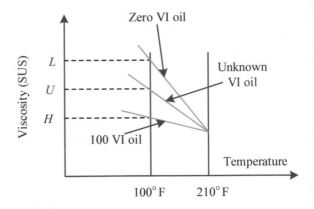

$L$ is the viscosity in SUS (Saybolt Universal Second) of 0-VI oil at 100 °F,
$U$ is the viscosity in SUS of unknown-VI oil at 100 °F,
$H$ is the viscosity in SUS of 100-VI oil at 100 °F.

*Example 3.9* The viscosity index of a sample oil of 65. It is tested with a 0 VI oil and a 100 VI oil whose viscosity values at 100 °F are 356 and 121 SUS, respectively. Find the viscosity of the sample oil at 100 °F in units of SUS.

**Solution**
The value of the U is calculated as,

$$65 = \frac{356 - U}{356 - 121} \times 100 \qquad (3.39)$$

$$152.75 = 356 - U \qquad (3.40)$$

$$U = 203.25 \, \text{SUS} \qquad (3.41)$$

**Practice Problem 3.9**
Find the viscosity index of sample oil is tested with a 0 VI oil and a 100 VI oil whose viscosity values at 42 °F are 425 and 139 SUS, respectively. The viscosity of the sample oil at 42 °F is 215 SUS.

## 3.10   Fluid Flow and Continuity Equation

The amount of flow of fluid per second is very important to accomplish as it is used to run the cylinders or actuators. Let us consider that the fluid enters into a pipe from

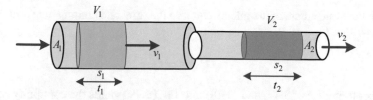

**Fig. 3.7** Fluid flow in a pipe

the left side and comes out from a pipe as shown in Fig. 3.7. The area, velocity, and volume at the input are represented as $A_1$, $v_1$, and $V_1$. The same parameters at the output are represented as, $A_2$, $v_2$, and $V_3$.

The velocity ($v$) is defined as distance ($s$) per unit time ($t$) and it is expressed as,

$$v = \frac{s}{t} \tag{3.42}$$

$$s = v \times t \tag{3.43}$$

The volume at the inlet of the pipe is expressed as,

$$V_1 = \text{area} \times \text{distance} = A_1 \times s_1 = A_1 \times v_1 t_1 \tag{3.44}$$

The volume at the output of the pipe is expressed as,

$$V_2 = \text{area} \times \text{distance} = A_2 \times s_2 = A_2 \times v_2 t_2 \tag{3.45}$$

The volume of fluid passing through a pipe per unit time is known as flow rate and it is denoted by $Q$. In an Imperial and US customer unit of the flow rate is expressed as gallons per minute (**GPM**). In the SI unit, the unit of flow rate is litter per minute (**LPM**). Mathematically, it is expressed as,

$$Q = \frac{V}{t} \tag{3.46}$$

Substituting Eq. (3.44) into Eq. (3.46) yields the flow rate at the inlet as,

$$Q_1 = \frac{A_1 s_1}{t} = A_1 v_1 \tag{3.47}$$

Similarly, the flow rate at the outlet of the pipe is,

$$Q_1 = \frac{V_2 s_2}{t} = A_2 v_2 \tag{3.48}$$

The flow rate enters at the inlet is equal to the flow rate coming out from the pipe and is written as,

$$Q_1 = Q_2 \tag{3.49}$$

Substituting Eqs. (3.47) and (3.48) into Eq. (3.49) yields the continuity equation as,

$$A_1 v_1 = A_2 v_2 \tag{3.50}$$

Alternative approach: The continuity equation can also be derived from the basic definition of density as,

$$m = \rho V \tag{3.51}$$

Differentiating both sides of Eq. (3.51) yields,

$$d(m) = d(\rho V) \tag{3.52}$$

Substituting $V = As$ into Eq. (3.52) yields,

$$dm = \rho dV = \rho d(A \times s) \tag{3.53}$$

$$dm = \rho dV = \rho A ds \tag{3.54}$$

The mass flow rate is defined as,

$$\dot{m} = \frac{dm}{dt} \tag{3.55}$$

Substituting Eq. (3.54) into Eq. (3.55) yields,

$$\dot{m} = \frac{dm}{dt} = \rho A \frac{ds}{dt} \tag{3.56}$$

Substituting, the velocity, $v = ds/dt$ in Eq. (3.56) yields,

$$\dot{m} = \rho A v \tag{3.57}$$

Due to mass conservation, the mass flow rate in the input and output of a pipe will be the same and it can be expressed as,

$$\dot{m}_1 = \dot{m}_2 \, \rho_1 A_1 v_1 = \rho_2 A_2 v_2 \tag{3.58}$$

If the fluid is incompressible, then $\rho_1 = \rho_2$, Eq. (3.58) can be expressed as,

$$A_1 v_1 = A_2 v_2 \tag{3.59}$$

Equation (3.59) is known as the equation of continuity or continuity equation.

***Example 3.10*** Find the flow velocity in m/s if the fluid flows at a rate of 25 LPM through a pipe with a diameter of 25 mm.

**Solution**
The value of the area calculated as,

$$A = \frac{\pi D^2}{4} = \frac{\pi \left(\frac{25}{1000}\right)^2}{4} = 4.91 \times 10^{-4} \, \text{m}^2 \tag{3.60}$$

The flow rate is converted to m$^3$/s as,

$$Q = \frac{25 \times 0.001}{60} = 4.17 \times 10^{-4} \, \text{m}^3/\text{s} \tag{3.61}$$

The velocity of the fluid is calculated as,

$$v = \frac{Q}{A} = \frac{4.17 \times 10^{-4}}{4.91 \times 10^{-4}} = 0.85 \, \text{m/s} \tag{3.62}$$

***Example 3.11*** The fluid flows through a pipe of section 1 at a velocity of 95 in./min. The pipe has two sections and the diameters of sections 1 and 2 are 3.5 in. and 2 in., respectively.
    Calculate the velocity at section 2 and the flow rate at GPM.

**Solution**
The value of the fluid velocity at section 2 is calculated as,

$$v_2 = \frac{A_1 v_1}{A_2} = \frac{2.5^2 \times 95}{2^2} = 148.44 \, \text{in./min} \tag{3.63}$$

The flow rate is calculated as,

$$Q = A_1 v_1 = \pi \times \frac{2^2}{4} \times 95 = 74.61 \, \text{in.}^3/\text{min} \tag{3.64}$$

Using the conversion rule 231 in.$^3$/min $= 1$ GPM yields,

$$Q = \frac{74.61}{231} = 0.32 \, \text{GPM} \tag{3.65}$$

**Practice Problem 3.10**

The flow rate for a fluid power system is given by 25 GPM. Calculate the pipe diameter if the is not exceeded 12 m/s.

**Practice Problem 3.11**

A pipe has two sections 1 and 3. The fluid flows through a pipe of section 1 at a velocity of 8 m/min. The diameters of sections 1 and 2 are 80 mm and 50 mm, respectively. Find the flow rate and velocity in section 3.

## 3.11    Bernoulli's Equation

Daniel Bernoulli (1700–1782) was a Dutch-born scientist who studied in Italy and eventually settled in Switzerland. In 1738, he published "Hydrodynamica", his study in fluid dynamics, or the study of how fluids behave when they are in motion. Bernoulli stated in "Hydrodynamica" that as a fluid moves faster, it produces less pressure, whereas slower-moving fluids produce more pressure.

Bernoulli's equation states that where the velocity is low, the pressure is high and where the velocity is high, the pressure is low.

Consider some assumptions such as fluid is incompressible, laminar flow, no viscosity, constant density and volume to derive Bernoulli's equation. Let us consider the force $F_1$ is applied in the inlet of a pipe to push the fluid and it travels the distance $x_1$ as shown in Fig. 3.8. It also travels a distance $x_1$ at the output side of the pipe.

The kinetic energy at the inlet and the outlet of the pipe is,

$$KE_1 = \frac{1}{2}mv_1^2 \tag{3.66}$$

$$KE_2 = \frac{1}{2}mv_2^2 \tag{3.67}$$

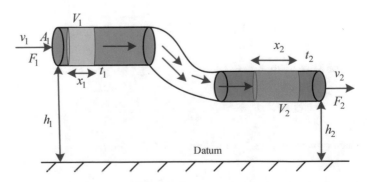

**Fig. 3.8**  A sample pipe for Bernoulli's equation

An increase in kinetic energy from inlet to outlet of a pipe as $v_2 > v_1$ is,

$$IKE = \frac{1}{2}mv_2^2 - \frac{1}{2}mv_1^2 \qquad (3.68)$$

The potential energy in the inlet and outlet of a pipe is,

$$PE_1 = mgh_1 \qquad (3.69)$$

$$PE_2 = mgh_2 \qquad (3.70)$$

The decrease in potential energy is,

$$DPE = mgh_1 - mgh_2 \qquad (3.71)$$

The work done in the inlet of the pipe is,

$$W_1 = F_1 x_1 = p_1 A_1 x_1 \qquad (3.72)$$

Similarly, the work done in the outlet of the pipe is,

$$W_2 = F_2 x_2 = p_2 A_2 x_2 \qquad (3.73)$$

The network done is calculated as,

$$\Delta W = W_1 - W_2 \qquad (3.74)$$

Substituting Eqs. (3.72) and (3.73) into Eq. (3.74) yields,

$$\Delta W = p_1 A_1 x_1 - p_2 A_2 x_2 \qquad (3.75)$$

Due to the conservation of energy, the increase in kinetic energy is equal to the sum of the decrease in potential energy and the network done. Mathematically, it can be expressed as,

$$IKE = DPE + \Delta W \qquad (3.76)$$

Substituting Eqs. (3.59) and (3.60) and expressions of $PE$ and $KE$ into Eq. (3.67) yields,

$$\frac{1}{2}mv_2^2 - \frac{1}{2}mv_1^2 = mgh_1 - mgh_2 + p_1 A_1 x_1 - p_2 A_2 x_2 \qquad (3.77)$$

Substituting the volume, $V = Ax$ and the mass $m = \rho V$ into Eq. (3.77)

$$\frac{1}{2}\rho V_2 v_2^2 - \frac{1}{2}\rho V_1 v_1^2 = \rho V_1 g h_1 - \rho V_2 g h_2 + p_1 V_1 - p_2 V_2 \qquad (3.78)$$

For constant velocity, considering $V_1 = V_2 = V$ into Eq. (3.78) yields,

$$\frac{1}{2}\rho v_2^2 - \frac{1}{2}\rho v_1^2 = \rho g h_1 - \rho g h_2 + p_1 - p_2 \qquad (3.79)$$

$$p_1 + \rho g h_1 + \frac{1}{2}\rho v_1^2 = p_2 + \rho g h_2 + \frac{1}{2}\rho v_2^2 \qquad (3.80)$$

Equation (3.80) is known as Bernoulli's equation in pressure form and it is expressed as,

$$p \text{ (static pressure)} + \rho g h \text{ (hydrostatic pressure)} + \frac{1}{2}\rho v^2 \text{ (dynamic pressure)}$$

$$= \text{constant} \qquad (3.81)$$

Dividing Eq. (3.80) by $\rho g$ yields,

$$\frac{p_1}{\rho g} + \frac{\rho g h_1}{\rho g} + \frac{\frac{1}{2}\rho v_1^2}{\rho g} = \frac{p_2}{\rho g} + \frac{\rho g h_2}{\rho g} + \frac{\frac{1}{2}\rho v_2^2}{\rho g} \qquad (3.82)$$

$$h_1 + \frac{p_1}{\rho g} + \frac{1}{2g}v_1^2 = h_2 + \frac{p_2}{\rho g} + \frac{1}{2g}v_2^2 \qquad (3.83)$$

Again, substituting $\gamma = \rho g$ in Eq. (3.83) yields,

$$h_1 + \frac{p_1}{\gamma} + \frac{1}{2g}v_1^2 = h_2 + \frac{p_2}{\gamma} + \frac{1}{2g}v_2^2 \qquad (3.84)$$

Equation (3.84) is known as Bernoulli's equation in head form and it is expressed as,

$$h \text{ (potential head)} + \frac{p_1}{\gamma} \text{ (pressure head)} + \frac{1}{2g}v^2 \text{ (kinetic head)} = \text{constant} \quad (3.85)$$

Multiplying Eq. (3.83) by $g$ yields,

$$g h_1 + \frac{p_1}{\rho} + \frac{1}{2}v_1^2 = g h_2 + \frac{p_2}{\rho} + \frac{1}{2}v_2^2 \qquad (3.86)$$

Equation (3.86) is known as Bernoulli's equation in energy form and is expressed as,

$$gh \text{ (potential energy)} + \frac{p_1}{\rho} \text{ (pressure energy)} + \frac{1}{2}v^2 \text{ (kinetic energy)} = \text{constant}$$
$$\tag{3.87}$$

Assume that there is no change in height of the fluid, then the term $\rho gh$ can be cancelled from each side. Equation (3.80) is revised as,

$$p_1 + \frac{1}{2}\rho v_1^2 = p_2 + \frac{1}{2}\rho v_2^2 \tag{3.88}$$

**Example 3.12** A fluid with a $\gamma = 8600$ N/m$^3$ flows through a system at a constant flow rate of 15 lpm. The areas at the input and output are equal. Calculate the pressure at the output terminals if the pressure at the input terminal is 650 kPa and the height between the outlet and inlet is $h = h_2 - h_1 = 12$ m.

**Solution**
The flow rate is constant so that the velocity at the input and output are the same and the Bernoulli's equation reduces to,

$$p_1 + \rho g h_1 = p_2 + \rho g h_2 \tag{3.89}$$

Substituting relevant values in Eq. (3.89) yields,

$$p_2 = p_1 + \gamma(h_1 - h_2) = p_1 - \gamma(h_2 - h_1) = 650{,}000 - 8600 \times 12 \tag{3.90}$$

$$p_2 = 546{,}800 \text{ Pa} = 546.8 \text{ kPa} \tag{3.91}$$

**Example 3.13** A fluid with a $\gamma = 0.05$ lb/in.$^3$ flows through a system at a constant flow rate of 250 in.$^3$/s. The areas are $A_1 = 3.5$ in.$^2$ and $A_2 = 1.5$ in.$^3$ Find the pressure at point 1, if the pressure at point 2 is 145 psi and the heights are equal.

**Solution**
The values of flow velocities are calculated as,

$$v_1 = \frac{Q_1}{A_1} = \frac{250}{2.5} = 100 \text{ in./s} \tag{3.92}$$

$$v_2 = \frac{Q_2}{A_2} = \frac{250}{1.5} = 166.67 \text{ in./s} \tag{3.93}$$

For equal height, Bernoulli's equation is written as,

$$\frac{p_1}{\gamma} + \frac{1}{2g}v_1^2 = \frac{p_2}{\gamma} + \frac{1}{2g}v_2^2 \tag{3.94}$$

Substituting relevant parameters in Eq. (3.94) yields,

$$p_1 = 145 + \frac{0.05}{2 \times 32.2 \times 12}\left(166.67^2 - 100^2\right) \tag{3.95}$$

$$p_1 = 146.15\,\text{psi} \tag{3.96}$$

**Practice Problem 3.12**

A fluid with a $\gamma = 9800$ N/m³ flows through a system at a constant flow rate of 12 LPM. The areas at the input and output are equal. Calculate the pressure at the output terminals if the pressure at the input terminal is 700 kPa and heights $h_1 = 12$ m and $h_2 = 20$ m.

**Practice Problem 3.13**

A fluid with a $\gamma = 0.03$ lb/in.³ flows through a system at a constant flow rate of 450 in.³/s. The areas are $A_1 = 3.5$ in.² and $A_2 = 3.5$ in.³· Find the pressure at point 1, if the pressure at point 1 is 150 psi and the heights are equal.

## 3.12   Torricelli's Theorem

Evangelista Torricelli (1608–1647) was an Italian physicist and mathematician who invented the barometer and whose work in geometry aided in the eventual development of integral calculus. Torricelli included his findings on fluid motion and projectile motion. Torricelli's theorem is a special case of Bernoulli's equation that explains the relationship between fluid leaving a hole and the liquid's height in that container. Let us consider an open container filled with a liquid as shown in Fig. 3.9.

At point 1, the pressure $p_1$ is atmospheric pressure and the velocity of fluid $v_1 = 0$. The same parameter at point 2 is $p_2$ is atmospheric pressure and the velocity $v_2$ is very high as it is coming out through a narrow pipe. The flow rates of the input and output can be expressed as,

**Fig. 3.9** Liquid with a container

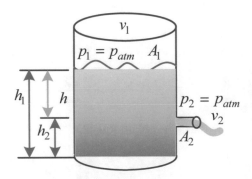

$$A_1 v_1 = A_2 v_2 \tag{3.97}$$

$$v_1 = \frac{A_2 v_2}{A_1} \tag{3.98}$$

Area $A_1$ is very higher than the area $A_2$, then the term $A_2 v_2$ is very low and Eq. (3.98) is modified as,

$$v_1 = \frac{A_2 v_2}{A_1} \approx 0 \, \text{m/s} \tag{3.99}$$

For the points 1 and 2, Bernoulli's equation can be written as,

$$p_1 + \frac{1}{2}\rho v_1^2 + \rho g h_1 = p_2 + \frac{1}{2}\rho v_2^2 + \rho g h_2 \tag{3.100}$$

Substituting Eq. (3.99) into Eq. (3.100) yields,

$$P_{atm} + \rho g h_1 = P_{atm} + \frac{1}{2}\rho v_2^2 + \rho g h_2 \tag{3.101}$$

$$\rho g h_1 - \rho g h_2 = \frac{1}{2}\rho v_2^2 \tag{3.102}$$

$$v_2 = \sqrt{g(h_1 - h_2)} \tag{3.103}$$

If $h = h_1 - h_2$, then Eq. (3.103) is modified as,

$$v_2 = \sqrt{2gh} \tag{3.104}$$

From Eq. (3.104), it is seen that the velocity of the discharge fluid depends on the height of the fluid.

***Example 3.14*** A tank is filled with fluid to a height of 3 ft. If the fluid leaks from the tank, calculate the velocity at the outlet.

**Solution**
The value of the discharge velocity is calculated as,

$$v_2 = \sqrt{2gh} = \sqrt{2 \times 32.2 \times 3} = 13.9 \, \text{ft/s} \tag{3.105}$$

**Practice Problem 3.14**
A tank is filled with a liquid to a specified height. If the fluid comes out from the tank with a velocity of 12 m/s, calculate the height of the liquid.

## 3.13  Young's Modulus

Consider a container full of water and place a sphere inside the container as shown in Fig. 3.10. The fluid will exert a force in all directions on the sphere and the volume of the sphere will change (will decrease) as can be seen in Fig. 3.11. Let the initial volume of the sphere is $V_0$ and the final volume is $V_f$.

The change in volume is expressed as,

$$-\Delta V = V_f - V_0 \tag{3.106}$$

As the volume decrease, the change in volume is negative. A force is applied horizontally to an object that is compressed as shown in Fig. 3.12. A force is also applied to an object and it is extended. The elongation per unit length is called the strain and it is denoted by $\varepsilon$. The strain is expressed as,

$$\varepsilon = \frac{\Delta L}{L} \tag{3.107}$$

Normal stress $\sigma$ is expressed as,

**Fig. 3.10**  A container with a sphere

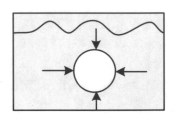

**Fig. 3.11**  Different sphere volumes

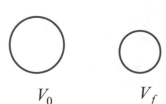

**Fig. 3.12**  Compression and extension of an object

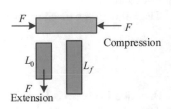

$$\sigma = \frac{P}{A} \tag{3.108}$$

The ratio of the unit stress to the unit strain is the modulus of elasticity of the material in tension, or, as it is often called, Young's modulus, $E$ is expressed as,

$$E = \frac{\sigma}{\varepsilon} \tag{3.109}$$

Substituting Eqs. (3.107) and (3.109) into Eq. (3.98) yields,

$$E = \frac{\frac{P}{A}}{\frac{\Delta L}{L}} \tag{3.110}$$

$$E = \frac{P}{A} \times \frac{L}{\Delta L} \tag{3.111}$$

From Eq. (3.111), it is seen that Young Modulus can be calculated if other parameters are known.

## 3.14 Bulk Modulus

Consider the fluid with a particular behaviour to explain the Bulk modulus. The volume of this type of fluid for a given mass can easily be changed when there is a change in pressure. This type of special property represents the compressibility of a fluid. Therefore, the Bulk modulus is defined as the property of the fluid that is commonly used to characterize compressibility. Bulk modulus is a numerical constant that describes the elastic properties of a solid or fluid when it is under pressure on all surfaces. A cubic and sphere shape with an applied force is shown in Fig. 3.13. Let the pressure $p$ be applied to the fluid and its initial volume is $V$. Then the pressure increases to $p + dp$ and the volume of the fluid decreases to $V - dV$. It is also defined as the ratio of pressure applied to the fractional change in volume.

**Fig. 3.13** Different shapes with forces

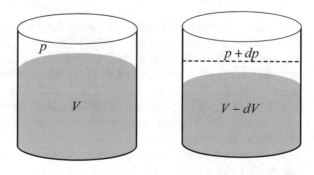

Mathematical, the bulk modulus, $B$ is expressed as,

$$B = \frac{\text{Bulk stress}}{\text{Bulk strain}} = \frac{\text{Volumetric stress}}{\text{Volumteric strain}} = \frac{p + dp - p}{\frac{V - dV - V}{V}} = \frac{dp}{\frac{-dV}{V}} = \frac{\Delta p}{-\Delta V / V}$$

(3.112)

where,

$B$ is the bulk modulus in the units of psi, Pa,
$\Delta p$ is the change in pressure in the units of psi, Pa,
$\Delta V$ is the change in volume in the units of in.$^3$, m$^3$,
$V$ is the initial volume of the fluid in the units of in.$^3$, m$^3$.

**Example 3.15** A 10 in$^3$ sample of oil is compressed in a cylinder until its pressure is increased from 100 to 2000 psi. If the bulk modulus equals 250,000 psi, find the change in volume in oil.


**Solution**
The value of the change in volume is calculated as,

$$\Delta V = -\frac{\Delta p}{B} V = -\frac{2000 - 100}{250{,}000} = -0.076 \, \text{in.}^3$$

(3.113)

This represents only a 0.76% decrease in volume, which shows that oil is highly incompressible.

**Practice Problem 3.15**
A 500 cm$^3$ sample of oil is to be compressed in a cylinder until its pressure is increased from 1 to 50 atm. If the bulk modulus of oil equals 1750 MPa, find the percentage change in its volume.


## 3.15   Reynold Number

In 1883, British engineer Professor Osborne Reynolds conducted an experiment to demonstrate the existence of three types of flow i.e. Laminar flow, transition flow and Turbulent flow. The type of flow in which the fluid molecules move along well-defined paths or all streamlines and all the streamlines are straight lines and parallel to the boundaries of the tube is known as laminar flow as shown in Fig. 3.14. Laminar flow depends on the high viscosity of the fluid, low velocity of the fluid and less area of the tube or pipe. The flow through a uniform cross-section of a pipe is an example of laminar flow in Fig. 3.14a.

The type of flow that the fluid molecules are way and parallel among themselves but are not parallel to the boundaries of the tube is known as transition flow as shown

**Fig. 3.14** Different types of flow

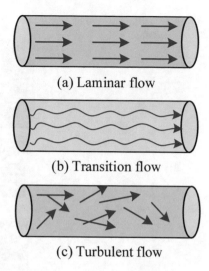

(a) Laminar flow

(b) Transition flow

(c) Turbulent flow

in Fig. 3.14b. The type of flow that the fluid molecules move in a zigzag way that led to the mixing of the dye is known as turbulent flow in Fig. 3.14c. In this type of flow, the fluid particles cross each other. The flow in a river during the flood and through a pipe of different cross-sections are examples of turbulent flow.

He developed a relationship to find the fluid friction in a hydraulic which is known as Reynold's number. Reynold's conducted an experiment related to fluid friction as shown in Fig. 3.15. At a low velocity, the dye will move in parallel to the tube boundaries and it also does not get any sort of dispersed as shown in Fig. 3.15a. This condition of the flow describes the laminar flow of the fluid. The velocity is slightly higher than the first case the dye will move in the way fashions as shown in Fig. 3.15b. This situation is expressed as the transition flow.

Whereas the velocity is higher than the second case, the dye will no longer move in the straight lines but will be diffused. This condition of the flow is usually described by the turbulent flow.

The ratio of inertia force to viscous force is called the Reynolds number ($R_e$) and mathematically, it is expressed as,

$$R_e = \frac{\text{Intertia force}}{\text{Viscous force}} = \frac{F_i}{F_v} \tag{3.114}$$

The resistance of an object (body) to a change in its state of motion is known as inertia force and it is expressed as,

$$F_i = ma \tag{3.115}$$

Substituting Eq. (3.5) into Eq. (3.115) yields,

$$F_i = \rho V a \tag{3.116}$$

**Fig. 3.15** Different types of flow filament

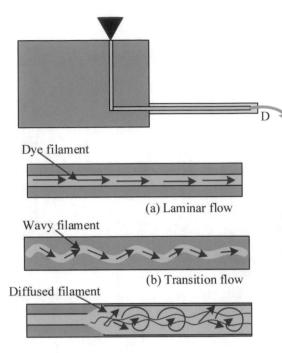

Dye filament

(a) Laminar flow

Wavy filament

(b) Transition flow

Diffused filament

Substituting the volume $V = A\,l$ in Eq. (3.116) yields,

$$F_i = \rho A l a \qquad (3.117)$$

Again, substituting $la = \text{m. m/s}^2 = (\text{m/s})^2 = v^2$ in Eq. (3.117) yields,

$$F_i = \rho A v^2 \qquad (3.118)$$

The resistance of a liquid to change of form is known as viscous force and it is expressed from Eq. (3.33) as,

$$F_v = \frac{\mu v A}{y} \qquad (3.119)$$

The distance $y$ is changed by the pipe diameter and Eq. (3.119) is revised as,

$$F_v = \frac{\mu v A}{D} \qquad (3.120)$$

Substituting Eqs. (3.120) and (3.118) into Eq. (3.114) yields,

$$R_e = \frac{F_i}{F_v} = \frac{\rho A v^2}{\frac{\mu v A}{D}} \qquad (3.121)$$

$$R_e = \frac{\rho v D}{\mu} \tag{3.122}$$

Equation (3.122) can be rearranged as,

$$R_e = \frac{v D}{\frac{\mu}{\rho}} \tag{3.123}$$

Substituting Eq. (3.34) into Eq. (3.123) yields,

$$R_e = \frac{v D}{v(nu)} \tag{3.124}$$

**Alternative approach**: Let us consider a fluid with a density of $\rho$ is flowing through a pipe whose length is $L$ as shown in Fig. 3.16. The velocity of a fluid is directly proportional to the fluid density, viscosity and length of the pipe. Mathematically, it is expressed as,

$$v \infty \rho^a \mu^b L^c \tag{3.125}$$

$$v = \text{Re}\rho^a \mu^b L^c \tag{3.126}$$

where Re is the Reynold number, $a$, $b$ and $c$ are the constants.

The dimension of fluid density, viscosity, velocity and length are derived as,

$$\rho = \frac{m}{V} \text{ kg/m}^3 = M^1 L^{-3} s^0 \tag{3.127}$$

$$\mu = \text{Pa.s} = \frac{N}{m^2}.s = \frac{kg.m}{s^2} \frac{1}{m^2}.s = \frac{kg}{s} \frac{1}{m} = M L^{-1} T^{-1} \tag{3.128}$$

$$v = \frac{m}{s} = M^0 L^1 T^{-1} \tag{3.129}$$

**Fig. 3.16** Different shapes with forces

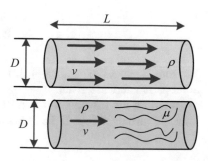

$$L = \text{m} = M^1 L^0 T^0 \tag{3.130}$$

Taking dimension both sides of Eq. (3.126) yields,

$$\dim(v) = \dim(\text{Re}\rho^a \, \mu^b \, L^c) \tag{3.131}$$

$$\dim(v) = \{\dim(\rho)\}^a \{\dim(\mu)\}^b \, \{\dim(L)\}^c \tag{3.132}$$

Substituting equations from (3.127) to (3.130) into Eq. (3.132) yields,

$$(M^0 L^1 T^{-1}) = \{M^1 L^{-3} s^0\}^a \{M^1 L^{-1} T^{-1}\}^b \, \{M^0 \, L^1 T^0\}^c \tag{3.133}$$

$$(M^0 L^1 T^{-1}) = M^{a+b} L^{-3a-b+c} \, T^{-b} \tag{3.134}$$

Equating coefficients of $M$, $L$ and $T$ yields,

$$a + b = 0 \tag{3.135}$$

$$-3a - b + c = 1 \tag{3.136}$$

$$b = 1 \tag{3.137}$$

Substituting Eq. (3.137) into Eq. (3.135) yields,

$$a = -1 \tag{3.138}$$

Substituting Eqs. (3.137), (3.138) into Eq. (3.136) yields,

$$c = -1 \tag{3.139}$$

Substituting Eqs. (3.137), (3.138) and (3.139) into Eq. (3.126) yields,

$$v = \text{Re}\rho^{-1} \, \mu^1 \, L^{-1} \tag{3.140}$$

$$\text{Re} = \frac{\rho L v}{\mu} \tag{3.141}$$

From Eqs. (3.124) and (3.141), it is seen that the Reynold can be calculated if other parameters are given.

Based on the Reynold number, the fluid is classified as laminar flow, transition flow and turbulent flow. If the Reynold number is less than 2200 then the fluid flow is known as laminar flow. The laminar appears when dealing with a small length of pipe and low velocity.

**Example 3.16** A fluid is flowing through a 1.5 in. diameter pipe with a kinematic viscosity of 0.0004 ft$^2$/s at a flow rate of 155 GPM. Find the Reynold number and identify the flow characteristics.

**Solution**
The value of the area is calculated as,

$$A = \frac{\pi D^2}{4} = \frac{\pi \times 1.5^2}{4} = 1.77 \, \text{in.}^2 \qquad (3.142)$$

The flow rate is converted as,

$$Q = \frac{155 \times 231}{60} = 596.75 \, \text{in.}^3/\text{s} \qquad (3.143)$$

The value of the velocity is determined as,

$$v = \frac{Q}{A} = \frac{596.75}{1.77} = 337.15 \, \text{in./s} = \frac{337.15}{12} = 28.09 \, \text{ft/s} \qquad (3.144)$$

The Reynold number is calculated as,

$$R_e = \frac{vD}{v(nu)} = \frac{28.09 \times 1.5/12}{0.0004} = 8779.86 \qquad (3.145)$$

According to Reynold's number, it is a laminar flow.

**Example 3.17** The viscosity and the relative density of fluid are 0.04 Ns/m$^2$ and 600 kg/m$^3$, respectively. The fluid flows through a 12 mm pipe with a velocity of 5 m/s. Determine the Reynold number.

**Solution**
The value of the area is calculated as,

$$Re = \frac{600 \times \frac{12}{1000} \times 5}{0.04} = 900 \qquad (3.146)$$

**Practice Problem 3.16**
A fluid is flowing through a 3.5 cm diameter pipe with a kinematic viscosity of 0.00224 m$^2$/s at a flow rate of 956 LPM. Find the Reynold number and identify the flow characteristics.

**Practice Problem 3.17**

The viscosity and the relative density of fluid are 0.002 Ns/m$^2$ and 1600 kg/m$^3$, respectively. The fluid flows through a 12 cm pipe with a velocity of 6 m/s. Find the Reynold number.

## 3.16  Buoyancy

Buoyancy is often known as the buoyant force. The buoyant force is the upward force on an object exerted by the surrounding fluid. When an object pushes water, the water pushes back with as much force as it can. If the water can push back as hard, the object floats such as a boat or fish. If not, it sinks such as steel. In other words, if the weight of an object is less than the Buoyant force, it will float. if the weight of an object is higher than the Buoyant force, it will sink. Buoyancy is caused by differences in pressure or force acting on opposite sides of an object immersed in a static fluid. The buoyancy is denoted by $B$ or $F_B$ and as a force, its unit is the newton (N). A boat and a weighted object are floated in the water of a container as shown in Fig. 3.17.

A cylindrical object is immersed inside a container as shown in Fig. 3.18. Let the height of the upper part and the lower part of the object from the top of the liquid

**Fig. 3.17**  A fish and boat in the water

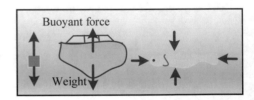

**Fig. 3.18**  An object inside a container

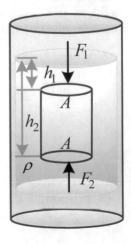

are $h_1$ and $h_2$, respectively. The force applied to the top and bottom of an object is $F_1$ and $F_2$, respectively.

The bottom of an object is deeper, so there is more force pushing up on it, which is known as a buoyant force. The buoyant force is the difference between the bottom and to forces and it can be expressed as,

$$F_B = F_2 - F_1 \qquad (3.147)$$

From the basic definition of pressure, the top and bottom forces are expressed as,

$$F_1 = P_1 A = \rho g h_1 A \qquad (3.148)$$

$$F_2 = P_2 A = \rho g h_2 A \qquad (3.149)$$

Substituting Eqs. (3.148) and (3.149) into Eq. (3.147) yields,

$$F_B = \rho g h_2 A - \rho g h_1 A = \rho g A (h_2 - h_1) \qquad (3.150)$$

$$F_B = \rho g V_d \qquad (3.151)$$

Again, consider the volume of an object is $V$. If the $V_d$ is greater than $V$, $F_B$ will be higher. As a result, the object will float. For a lower displacement volume, the buoyant force will be lower, and the object will sink. The actual mass and volume of the object are $m$ and $V$, respectively when measured in the air as shown in Fig. 3.19. Here, the weight of the object is,

$$T = W = mg = \rho_0 V_0 g \qquad (3.152)$$

When the object is immersed in the fluid, its weight will be reduced which is known as apparent weight, $W_a$. The net force around the object under immerged condition is,

**Fig. 3.19** An object inside and outside of a container

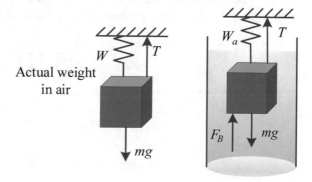

Actual weight in air

$$\sum F_{net} = T + F_B + mg = 0 \tag{3.153}$$

If 'up' is defined arbitrarily to be the 'positive' direction and 'down' is treated as being the 'negative' direction, then Eq. (3.153) is represented as,

$$T + F_B - mg = 0 \tag{3.154}$$

$$T = mg - F_B = \text{Apparent weight} (W_a) \tag{3.155}$$

Under the immersed condition, the Buoyancy force is expressed as,

$$F_B = W_{displaced\,liq} = (mg)_{\text{displaced liq}} = \rho_f V_d g \tag{3.156}$$

Substituting Eq. (3.152) into Eq. (3.155) yields,

$$T = W - F_B = \text{Apparent weight} (W_a) \tag{3.157}$$

$$F_B = W - W_a \tag{3.158}$$

Substituting Eq. (3.156) into Eq. (3.158) yields,

$$F_B = W - W_a = \rho_f V_d g \tag{3.159}$$

Dividing Eq. (3.152) by Eq. (3.159) yields,

$$\frac{W}{F_B} = \frac{W}{W - W_a} = \frac{\rho_0 V_0 g}{\rho_f V_d g} \tag{3.160}$$

$$\frac{W}{F_B} = \frac{W}{W - W_a} = \frac{\rho_0 V_0}{\rho_f V_d} \tag{3.161}$$

As the volume of the raft decreases, the volume of the displaced fluid decreases which in turn decreases the Buoyant force. After the raft becomes completely submerged, the two volumes are equation, Eq. (3.161) becomes,

$$\frac{W}{F_B} = \frac{W}{W - W_a} = \frac{\rho_0}{\rho_f} \tag{3.162}$$

Equation (3.162) can be again revised as,

$$(W - W_a)\rho_0 = \rho_f W \tag{3.163}$$

$$(\rho_0 - \rho_f)W = \rho_0 W_a \tag{3.164}$$

$$(\rho_0 - \rho_f)mg = \rho_0 m_a g \tag{3.165}$$

$$m_a = m\left(1 - \frac{\rho_f}{\rho_0}\right) \tag{3.166}$$

$$\rho_f = \rho_0\left(1 - \frac{m_a}{m}\right) \tag{3.167}$$

*Example 3.18* A 1200 kg/m$^3$ iron immersed in water. The displaced volume of water is $2 \times 10^{-4}$ m$^3$. Calculate the Buoyant force.

**Solution**

The value of the Buoyant force is calculated as,

$$F_B = \rho V_d g = 1200 \times 2 \times 10^{-4} \times 9.81 = 2.35\,\text{N} \tag{3.168}$$

*Example 3.19* A 45 N copper immersed in water. The density of water is 1000 kg/m$^3$ and the density of copper is 8960 kg/m$^3$. Determine the Buoyant force.

**Solution**

The value of the Buoyant force is calculated as,

$$\frac{F_B}{W_{cu}} = \frac{\rho_f}{\rho_{cu}} = \frac{1000}{8960} \tag{3.169}$$

$$\frac{F_B}{45} = \frac{1000}{8960} \tag{3.170}$$

$$F_B = \frac{1000}{8960} \times 45 = 5.02\,\text{N} \tag{3.171}$$

**Practice Problem 3.18**

A basketball floats in a large container of water. The mass of the ball is 0.35 kg. Calculate the Buoyant force.

**Practice Problem 3.19**

An object is immersed in water and the ratio of Buoyant force to the weight of the object is 0.25. Determine the density of the object.

## 3.17   Pascal Law

French mathematician, physicist Blaise Pascal (1623–1662) experimented that when a force is applied to a confined liquid, the change in pressure is transmitted equally to all parts of the fluid as shown in Fig. 3.20. According to his name, it is known as Pascal law. This law stated that "a change in pressure applied to any part of the enclosed fluid is transmitted without reduction to every point of the fluid and the walls of the container". In an enclosed fluid, the molecules of the fluid are free to move about. Therefore, the molecules of the fluid transmit pressure to all parts of the fluid and the walls of the container. The pressure is defined as the force per unit area and mathematically, it is expressed as,

$$P = \frac{F}{A} \tag{3.172}$$

The theme of Eq. (3.172) can be more visualized using Fig. 3.21. From Fig. 3.21 and Eq. (3.172), the expressions of force and area can also be written for specific calculation purposes.

**Fig. 3.20**  A container with liquid

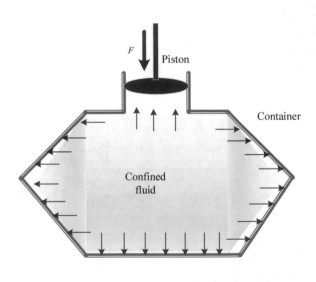

**Fig. 3.21**  A pressure triangle

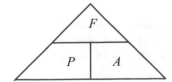

## 3.18   Force Transmission and Multiplication

Consider two same sizes cylinder with a confined fluid are connected through a pipe as shown in Fig. 3.22. A 10 N force is applied to the piston of the first cylinder and the area of each piston is $0.1\text{m}^3$. The pressure created at the first cylinder is calculated as,

$$p_1 = \frac{F_1}{A_1} = \frac{10}{0.1} = 100\,\text{N/m}^2 \tag{3.173}$$

According to Pascal law, pressure at a point acts in all directions and a change in pressure at any point in a fluid is equally transmitted throughout the entire fluid. Therefore, the pressure at the second cylinder is $100\,\text{N/m}^2$ and the force acted in the second cylinder is calculated as,

$$F_2 = p_1 A_2 = 100 \times 0.1 = 10\,\text{N} \tag{3.174}$$

It is seen that the 10 N force is applied to the input cylinder and the 10 N force comes out from the output or second cylinder. Therefore, the 10 N force is simply transmitted from the first cylinder to the second cylinder.

Two cylinders with different shapes are connected by the pipe as shown in Fig. 3.23. The area and the force for the first cylinder are $A_1$ and $F_1$. These parameters for the second cylinder are $A_2$ and $F_2$. A downward input force $F_1$ is applied to the smaller diameter piston of the first cylinder. This force produces an oil pressure $p_1$

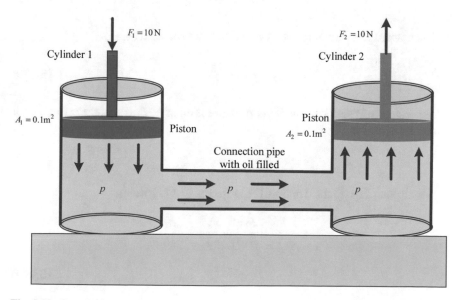

**Fig. 3.22**  Same cylinders for transmission of force

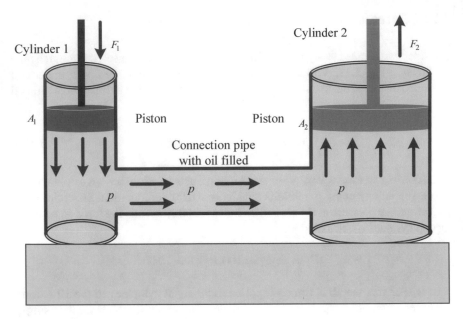

**Fig. 3.23** Different cylinders for multiplication of force

at the bottom of the first piston and mathematically, it can be expressed as,

$$p_1 = \frac{F_1}{A_1} \tag{3.175}$$

The upward pressure at the second cylinder is expressed as,

$$p_2 = \frac{F_2}{A_2} \tag{3.176}$$

According to Pascal law, the input pressure is equal to the output pressure and it can be expressed as,

$$p_1 = p_2 \tag{3.177}$$

Substituting Eqs. (3.175) and (3.176) into Eq. (3.177) yields,

$$\frac{F_1}{A_1} = \frac{F_2}{A_2} \tag{3.178}$$

$$F_2 = \frac{A_2}{A_1} F_1 \tag{3.179}$$

Let us consider 10 Pa pressure is applied to the input cylinder. The areas of the input and output cylinders are 1 m² and 4 m², respectively. The input force is calculated as,

$$F_1 = p_1 A_1 = 10 \times 1 = 10\,\text{N} \tag{3.180}$$

The output force is calculated as,

$$F_2 = \frac{A_2}{A_1} F_1 = \frac{4}{1} \times 10 = 40\,\text{N} \tag{3.181}$$

From Eq. (3.181), it is seen that the output force is multiplied by a factor of the area of the second cylinder.

The force $F_1$ is applied to the first cylinder that moves the piston to a distance $d_1$ and consequently moves the piston to the distance $d_2$ with the same amount of time as shown in Fig. 3.24. The work done by the first cylinder is expressed as,

$$W_1 = F_1 d_1 \tag{3.182}$$

The work done by the second cylinder is expressed as,

$$W_2 = F_2 d_2 \tag{3.183}$$

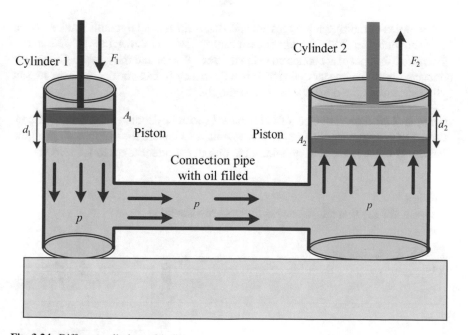

**Fig. 3.24** Different cylinders with distances

The work done by both cylinders are the same and it is expressed as,

$$W_1 = W_2 \tag{3.184}$$

Substituting Eqs. (3.182) and (3.183) into Eq. (3.184) yields,

$$F_1 d_1 = F_2 d_2 \tag{3.185}$$

$$p_1 A_1 d_1 = p_2 A_2 d_2 \tag{3.186}$$

Substituting Eq. (3.177) into Eq. (3.186) yields,

$$A_1 d_1 = A_2 d_2 \tag{3.187}$$

$$d_1 = \frac{A_2}{A_1} d_2 \tag{3.188}$$

The pistons of the cylinders move the distances with the same amount of time. Therefore, it can be replaced by the velocity ($v = d/t$) in Eq. (3.188) yields,

$$v_1 = \frac{A_2}{A_1} v_2 \tag{3.189}$$

Examples of multiplying forces are car lifts, forklifts and hydraulic brakes. A car lift is shown in Fig. 3.25. A hydraulic car brake system is shown in Fig. 3.26. In this system, the foot pedal is connected to the master cylinder and the wheel cylinders are connected with the master cylinder. When the driver brakes the car, then the master cylinder controls the other cylinder accordingly.

**Example 3.20** The diameters of the first and second cylinders are 1.5 in. and 2 in. A force of 160 lb is applied to the first cylinder. Calculate the output force and the distance to move in the first cylinder if the output cylinder moves to 1.5 in. distance.

**Solution**
The areas for the first and second cylinders are calculated as,

**Fig. 3.25** A schematic of car lift

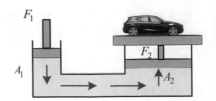

**Fig. 3.26** A hydraulic car brake system

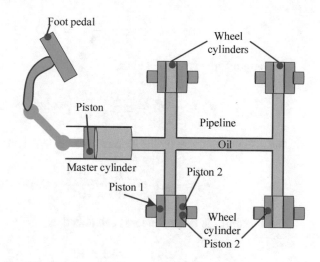

$$A_1 = \pi \frac{D_1^2}{4} = \pi \frac{1.5^2}{4} = 1.77 \, \text{in.}^2 \qquad (3.190)$$

$$A_2 = \pi \frac{D_2^2}{4} = \pi \frac{2^2}{4} = 3.14 \, \text{in.}^2 \qquad (3.191)$$

The output force is calculated as,

$$F_2 = \frac{A_2}{A_1} F_1 = \frac{3.14}{1.77} \times 160 = 283.84 \, \text{lb} \qquad (3.192)$$

The distance to move by the first cylinder is calculated as,

$$d_1 = \frac{A_2}{A_1} d_2 = \frac{3.14}{1.77} \times 1.5 = 2.66 \, \text{in.} \qquad (3.193)$$

***Example 3.21*** An input cylinder with a diameter of 30 mm is connected to an output cylinder with a diameter of 60 mm. A force of 500 N is applied to the first cylinder. Calculate the output force and the distance to move in the second cylinder if the input cylinder moves to a 10 mm distance. Also, calculate the speed of the input cylinder if the output cylinder moves at 5 m/s.

**Solution**
The areas for the first and second cylinders are calculated as,

$$A_1 = \pi \frac{D_1^2}{4} = \pi \frac{0.03^2}{4} = 7.07 \times 10^{-4} \, \text{m}^2 \qquad (3.194)$$

$$A_2 = \pi \frac{D_2^2}{4} = \pi \frac{0.06^2}{4} = 2.83 \times 10^{-3}\,\text{m}^2 \tag{3.195}$$

The output force is calculated as,

$$F_2 = \frac{A_2}{A_1} F_1 = \frac{2.83 \times 10^{-3}}{7.07 \times 10^{-4}} \times 500 = 2001.41\,\text{N} \tag{3.196}$$

The distance to move by the second cylinder is calculated as,

$$d_2 = \frac{A_1}{A_2} d_1 = \frac{7.07 \times 10^{-4}}{2.83 \times 10^{-3}} \times 0.01 = 2.50\,\text{mm} \tag{3.197}$$

The speed of the first cylinder is calculated as,

$$v_1 = \frac{A_2}{A_1} v_2 = \frac{2.83 \times 10^{-3}}{7.07 \times 10^{-4}} \times 5 = 20.01\,\text{m/s} \tag{3.198}$$

**Practice Problem 3.20**
The diameters of the first and second cylinders are 3.5 in. and 4 in. A force of 200 lb is applied to the first cylinder. Calculate the output force and the distance to move in the first cylinder if the output cylinder moves to 3.5 in. distance.

**Practice Problem 3.21**
An input cylinder with a diameter of 60 mm is connected to an output cylinder with a diameter of 80 mm. A force of 400 N is applied to the first cylinder. Calculate the output force and the distance to move in the second cylinder if the input cylinder moves to a 20 mm distance. Also, calculate the speed of the output cylinder if the input cylinder moves at 4 m/s.

## 3.19   Pressure Loss

Pressure drops usually occur when the fluid flows through a piping system. The pressure drops depend on the interior surface roughness ($e$), cross-sectional area ($A$) and length ($L$) of the pipe. In addition to that, it also depends on how many fittings and the geometrical complexity of each component are in the system. The higher the pressure drop in the line or piping system, the greater the amount of energy consumed to maintain the desired process flow and run the higher horsepower. The lower the pressure drop in a piping system, the less energy consumed, requiring a lower horsepower motor. Continuity equation, Bernoulli's equation, Reynolds number and Darcy's equation are required to calculate the pressure loss in a piping system.

## 3.20 Darcy-Weisbach Equation

In 1856, Henry Darcy, a French hydraulic engineer and Julius Weisbach, a German mathematician developed a mathematical model to calculate the pressure drop in a piping system that the fluid flow experiences due to the head loss. The frictional resistance in a pipe is independent of pressure, is proportional to the fluid flow ($Q$). The frictional resistance is directly proportional to the square of the fluid flow velocity i.e., above the critical velocity. The critical velocity is the velocity of a fluid at which the liquid turns from laminar to turbulent flow.

Consider a piece of pipe with a diameter $d$, the length between sections 1 and 2 is $L$. Let the pressure at the input is $p_1$ and at the output is $p_3$. Let the fluid enter the pipe at a velocity $v_1$ and comes out from the pipe at a velocity $v_3$. The fluid is getting frictional loss $f'$ due to the inner surface of the pipe. From Fig. 3.27, Bernoulli's equation in a head form including friction loss can be written as,

$$h_1 + \frac{p_1}{\rho g} + \frac{v_1^2}{2g} = h_2 + \frac{p_2}{\rho g} + \frac{v_2^2}{2g} + \left( \frac{\Delta p}{\rho g} = h_f \right) \tag{3.199}$$

The pipe is placed horizontally and the diameter is the same for sections 1 and 3. Therefore, $h_1 = h_2$ and $A_1$ and $A_2$ and Eq. (3.199) is modified as [3],

$$\frac{p_1}{\rho g} = \frac{p_2}{\rho g} + \left( \frac{\Delta p}{\rho g} = h_f \right) \tag{3.200}$$

$$h_f = \frac{p_1 - p_2}{\rho g} \tag{3.201}$$

The perimeter of the pipe is,

$$pe = \pi D \tag{3.202}$$

Let the frictional resistance per unit wetted area per unit velocity is $f'$. The frictional force ($F$) in the pipe due to turbulent fluid flow is expressed as,

$$F = f' \times \text{wetted area} \times (\text{velocity})^2 \tag{3.203}$$

**Fig. 3.27** A piece of pipe two sections

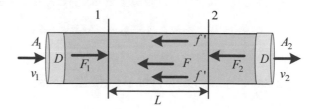

$$F = f' \times \pi DL \times (v)^2 \tag{3.204}$$

Substituting Eq. (3.202) into Eq. (3.204) yields,

$$F = f' \times pe \times L \times v^2 \tag{3.205}$$

The opposite force at section 1 of the pipe is,

$$F_1 = p_1 A \tag{3.206}$$

The opposite force at section 2 of the pipe is,

$$F_2 = p_2 A \tag{3.207}$$

Resolving forces in the horizontal direction and between the sections 1 and 2 yields,

$$F_1 - F_2 - F = 0 \tag{3.208}$$

Substituting Eqs. (3.206) and (3.207) into Eq. (3.208) yields,

$$p_1 A - p_2 A - F = 0 \tag{3.209}$$

$$\frac{F}{A} = p_1 - p_2 \tag{3.210}$$

Substituting Eq. (3.201) into Eq. (3.210) yields,

$$\frac{F}{A} = h_f \rho g \tag{3.211}$$

Substituting Eq. (3.204) into Eq. (3.211) yields,

$$\frac{f' \times pe \times L \times v^2}{A} = h_f \rho g \tag{3.212}$$

But, the perimeter over the area of the pipe is expressed as,

$$\frac{pe}{A} = \frac{\pi D}{\pi \frac{D^2}{4}} = \frac{4}{D} \tag{3.213}$$

Substituting Eq. (3.213) into Eq. (3.212) yields,

$$\frac{f' \times 4 \times L \times v^2}{D} = h_f \rho g \tag{3.214}$$

$$h_f = \frac{f'}{\rho} \times \frac{4 \times L \times v^2}{gD} \tag{3.215}$$

Again, considering $f'/\rho = f/2$, where $f$ is the coefficient of the friction or Darcy's friction factor. Equation (3.215) is revised as,

$$h_f = \frac{f}{2} \times \frac{4 \times L \times v^2}{gD} \tag{3.216}$$

In general, the number 4 is omitted to calculate the exact head loss and the Dary's-Weisbach equation is expressed as,

$$h_f = f \frac{L \times v^2}{2gD} \tag{3.217}$$

In 1937, Cyril Frank Colebrook and Cedric Masey White, British medical researchers introduced an equation to solve for the turbulent friction factor rather than use the Moody diagram (Lewis Ferry Moody in 1994). The Colebrook and White equation is,

$$\frac{1}{\sqrt{f}} = -2\log_{10}\left[\frac{e}{3.7D} + \frac{2.51}{\mathrm{Re}\sqrt{f}}\right] \tag{3.218}$$

Swamee-Jain's equation is also used to find the friction factor in a full-flowing circular pipe directly. It is an approximation of Colebrook and White's equation and this equation is,

$$f = \frac{0.25}{\left[\log_{10}\left(\frac{e}{3.7D} + \frac{5.74}{\mathrm{Re}^{0.9}}\right)\right]^2} \tag{3.219}$$

Serghides' solution was based on Steffensen's method. The solution involves the calculation of three intermediate values first and then substituting those values into a final equation. These steps are,

$$A = -2\log_{10}\left[\frac{e}{3.7D} + \frac{12}{\mathrm{Re}}\right] \tag{3.220}$$

$$B = -2\log_{10}\left[\frac{e}{3.7D} + \frac{2.51 \times A}{\mathrm{Re}}\right] \tag{3.221}$$

$$C = -2\log_{10}\left[\frac{e}{3.7D} + \frac{2.51 \times B}{\mathrm{Re}}\right] \tag{3.222}$$

$$f = \left[ A - \frac{(B - A)^2}{C - 2B + A} \right]^{-2}$$

(3.223)

The head loss in fittings can be determined using loss coefficients (k-Factors). In this approach, the k-factor is multiplied by the velocity head of the fluid flow as [4],

$$h_f = k \times \frac{v^2}{2g}$$

(3.224)

Substituting Eq. (3.217) into Eq. (3.224) yields,

$$k \times \frac{v^2}{2g} = f \frac{L}{D} \frac{v^2}{2g}$$

(3.225)

$$k = f \frac{L}{D}$$

(3.226)

In 1994, Lewis Ferry Moody developed and presented a diagram to calculate the friction factor for the piping system. According to his name, it is known as the Moody diagram as shown in Fig. 3.28.

***Example 3.22*** A 100 m cast-iron pipe carries water at 20 °C with a velocity of 2 m/s. The diameter of the pipe is 2 in and the pipe roughness is 0.15 mm. Calculate the frictional head loss of the pipe.

**Solution**
The diameter of the pipe is,

$$D = 2 \, \text{in.} = 2 \times 2.54 \, \text{cm} = 50.8 \, \text{mm}$$

(3.227)

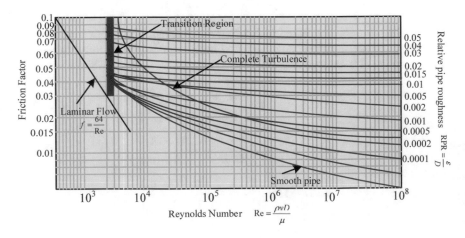

**Fig. 3.28** Moody diagram

The relative pipe roughness (RPR) is calculated as,

$$RPR = \frac{\varepsilon}{D} = \frac{0.15}{50.8} = 0.003 \qquad (3.228)$$

Based on the relative pipe roughness value, an approximate value of the friction factor from the Moody diagram is calculated as,

$$f = 0.028 \qquad (3.229)$$

The head loss of the pipe due to friction is calculated as,

$$h_f = f \frac{L \times v^2}{2gD} = 0.028 \times \frac{100 \times 2^2}{2 \times 9.81 \times 0.0508} = 11.24\,\text{m} \qquad (3.230)$$

**Practice Problem 3.22**

A 120 m cast-iron pipe carries water at 20 °C with a velocity of 3 m/s. The diameter of the pipe is 60 mm and the pipe roughness is 0.25 mm. Calculate the frictional head loss of the pipe.

**Exercise Problems**

3.1   The mass of an object is found to be 250 kg. Find its weight.

3.2   The weight of an object is found to be 500 lb. Calculate its mass.

3.3   A container is filled with a liquid whose weight is found to be 1000 pounds. The specification of a container is 4 ft long, 3.5 ft wide, and 5 ft deep. Find the density.

3.4   A cylinder container has a diameter of 0.25 m and a height of 3 m. If it is filled with a liquid having a specific weight of 1200 N/m³, how many kg of this liquid must be added to fill the container?

3.5   The density of gold is 21,500 kg/m³ and the density of water is 1000 kg/m³. Find the specific gravity.

3.6   The pressure at the bottom of an oil-filled tank is 12 kPa and the specific weight is 2200 N/m³. Determine the depth of the tank.

3.7   The absolute and density of fluid are given as 0.75 Ns/m² and 3.5 kg/m³. Calculate the kinematic viscosities of a fluid.

3.8   A 15 N force moves a piston inside a cylinder at a velocity of 3.5 m/s. The diameters of the cylinder and piston are 6 cm and 4.9 cm, respectively. An oil film separates the piston from the cylinder and the length of the piston is 6.2 cm. Calculate the absolute viscosity of the oil.

3.9   Find the viscosity index of sample oil is tested with a 0 VI oil and a 100 VI oil whose viscosity values at 42 °F are 435 and 143 SUS, respectively. The viscosity of the sample oil at 42 °F is 225 SUS.

3.10  The flow rate for a fluid power system is given by 35 GPM. Calculate the pipe diameter if the is not exceeded 9 m/s.

3.11    A pipe has two sections 1 and 3. The fluid flows through a pipe of section 1 at a velocity of 6 m/min. The diameters of sections 1 and 2 are 75 mm and 55 mm, respectively. Find the flow rate and velocity in section 3.

3.12    A fluid with a $\gamma = 9500$ N/m$^3$ flows through a system at a constant flow rate of 22 LPM. The areas at the input and output are equal. Calculate the pressure at the output terminals if the pressure at the input terminal is 740 kPa and heights $h_1 = 14$ m and $h_2 = 19$ m.

3.13    A fluid with a $\gamma = 0.04$ lb/in.$^3$ flows through a system at a constant flow rate of 550 in$^3$/s. The areas are $A_1 = 4.5$ in.$^2$ and $A_2 = 4.5$ in.$^3$. Find the pressure at point 1, if the pressure at point 1 is 155 psi and the heights are equal.

3.14    A tank is filled with a liquid to a specified height. If the fluid comes out from the tank with a velocity of 15 m/s, calculate the height of the liquid.

3.15    A 650 cm$^3$ sample of oil is to be compressed in a cylinder until its pressure is increased from 1 to 55 atm. If the bulk modulus of oil equals 1700 MPa, find the percentage change in its volume.

3.16    A fluid is flowing through a 4.5 cm diameter pipe with a kinematic viscosity of 0.00227 m$^2$/s at a flow rate of 900 LPM. Find the Reynold number and identify the flow characteristics.

3.17    The viscosity and the relative density of fluid are 0.004 Ns/m$^2$ and 1400 kg/m$^3$, respectively. The fluid flows through a 16 cm pipe with a velocity of 6 m/s. Find the Reynold number.

3.18    A basketball floats in a large container of water. The mass of the ball is 0.45 kg. Calculate the Buoyant force.

3.19    An object is immersed in water and the ratio of Buoyant force to the weight of the object is 0.35. Determine the density of the object.

3.20    The diameters of the first and second cylinders are 3.5 in. and 4.5 in., respectively. A force of 250 lb is applied to the first cylinder. Calculate the output force and the distance to move in the first cylinder if the output cylinder moves to 3.5 in. distance.

3.21    An input cylinder with a diameter of 65 mm is connected to an output cylinder with a diameter of 85 mm. A force of 450 N is applied to the first cylinder. Calculate the output force and the distance to move in the second cylinder if the input cylinder moves to a 25 mm distance. Also, calculate the speed of the output cylinder if the input cylinder moves at 3 m/s.

3.22    A 125 m cast-iron pipe carries water at 20 °C with a velocity of 4 m/s. The diameter of the pipe is 70 mm and the pipe roughness is 0.10 mm. Calculate the frictional head loss of the pipe.

# References

1.  J.L. Johnson, *Introduction to Fluid Power* (Delmar, USA, 2003), pp. 1–502
2.  F. Don Norvelle, *Fluid Power Technology* (Delmar, a Division of Thomson Learning, USA, 1995), pp. 1–649

3. F.M. White, *Fluid Mechanics*, 7th edn. (McGraw-Hill, USA, 2011), pp. 1–862
4. ASHARE, *Handbook Fundamentals*, I-P edn. (2017)

# Chapter 4
# Hydraulic Pumps

## 4.1 Introduction

A hydraulic pump is a very important device to run the hydraulic power systems smoothly. The pump is attached either with a single-phase or a three-phase induction motor. The pump shaft and the motor shaft are coupled with gasket, nuts and bolts. In a positive half of the cycle, the pump creates the vacuum which in turn increases the volume and pulls the water from the source such as a lake, pond and river. Whereas in the negative half of the cycle, it discharges the water by reducing the volume. The necessary pipe fittings and accessories are used to control water flow through the piping system. Regarding pumps, some important points need to know. The pump generates flow, not pressure. However, the pressure is the result of resistance to the fluid flow and the pressure will be maximum if the resistance to the fluid flow is minimum. This chapter will discuss pump classification, construction, working principles, power and torque expressions.

## 4.2 Pump

A pump, which is the heart of a hydraulic system, converts mechanical energy into hydraulic energy. The mechanical energy is delivered to the pump via a prime mover such as an electric motor. Due to mechanical action, the pump creates a partial vacuum at its inlet. This permits atmospheric pressure to force the fluid through the inlet line and into the pump. The pump then pushes the fluid into the hydraulic system. However, the pump does not generate pressure [1].

A pump is a hydraulic machine that is used to import hydraulic energy to water and or liquid in the form of pressure energy or kinetic energy. In other words, a pump is a device used to move fluids (liquids or gases) from one place to another place. A pump is a hydraulic machine that converts mechanical energy to hydraulic energy (pressure energy). A hydraulic pump takes fluid (oil or water) from a tank

Md. A. Salam, *Fundamentals of Pneumatics and Hydraulics*,
https://doi.org/10.1007/978-981-19-0855-2_4

and delivers it to the hydraulic distribution circuit. In doing so it raises the pressure of the fluid to the required level as shown in Fig. 4.1.

The pressure of the fluid at the pump outlet is zero if it is not connected to a system. This pressure rises from zero to a required level when it is connected to a system. A hydraulic pump speed depends on the three-phase induction motor shaft speed as shown in Fig. 4.2. The motor and pump must be rotated in a clockwise direction. Before coupling with a pump, the three-phase connection should be checked and tested the rotation. If the motor rotation is not clockwise direction, then any two supply lines need to be swapped.

The speed of a pump shaft depends on the motor shaft speed and the motor shaft speed is expressed as,

$$N_s = \frac{120f}{P}$$

(4.1)

where,

$f$ is the frequency in units of Hertz (Hz),
$P$ is the number of poles,
$N_s$ is the speed of the motor in units of revolution per minute (rpm).

**Example 4.1** The number of poles of a three-phase induction motor is 4. Find the speed of the pump if the frequency is 60 Hz.

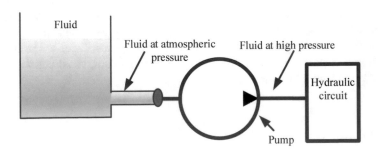

**Fig. 4.1** A fluid tank with a pump

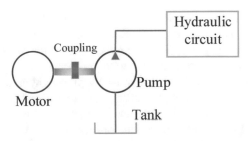

**Fig. 4.2** A fluid tank with a pump

**Solution**

The value of the pump speed is calculated as,

$$N_s = \frac{120f}{P} = \frac{120 \times 60}{4} = 1800 \, \text{rpm} \tag{4.2}$$

**Practice Problem 4.1**

The speed of the is found to be 1200 rpm. Calculate the number of poles if the frequency is 60 Hz.

## 4.3  Pump Pressure and Flow Rate

Pressure and flow rate are very important for understanding the working principle of the pump. The pressure at the suction pipe (inlet) is very less and at the delivery pipe is high. This pressure is directly proportional to the system's applied force. A pump is conned to two cylinders $a$ and $b$. The area and force of cylinder $a$ are 2 in.$^2$ and 1200 lb as shown in Fig. 4.1. The same parameters for cylinder $b$ are 3 in.$^2$ and 1500 lb. The pressure for cylinder $a$ is calculated as [2],

$$P_a = \frac{F_a}{A_a} = \frac{1200}{2} = 600 \, \text{psi} \tag{4.3}$$

The pressure for the second cylinder is calculated as,

$$P_B = \frac{F_B}{A_B} = \frac{1500}{3} = 500 \, \text{psi} \tag{4.4}$$

From Eqs. (4.3) and (4.4), it is seen that the pump will first reach cylinder B and performs the work. After this, the remaining pressure will reach cylinder B (Fig. 4.3).

The amount (volume) of fluid that delivers per unit time through the discharge pipe is known as flow rate. All pumps are connected to the external power source. The three-phase or single-phase induction motor is used as the external power source. The speed of the motor depends on the number of poles and frequency. Therefore, the pump flow rate is related to the displacement per revolution and the motor speed revolution per min (rpm). Mathematically, the theoretical pump flow rate is expressed as,

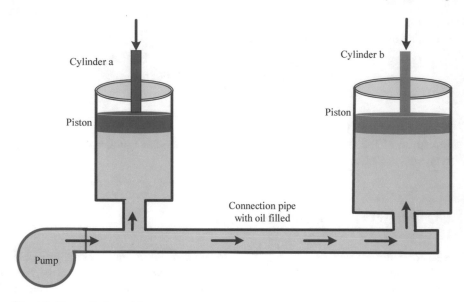

**Fig. 4.3** Two cylinders with pump

$$Q_T = V_P \times N \tag{4.5}$$

where,

$V_P$ is the pump displacement in the units of m³/rev, in.³/rev,
$N$ is the pump speed in the unit of rev/min,
$Q_T$ is the theoretical flow rate in the units of m³/min, in.³/min.

The theoretical pump flow rate in terms of a gallon per minute can be expressed using the factor 1 gallon equal to 231 in.³ as,

$$Q_T = \frac{V_P \times N}{231} \text{ gpm} \tag{4.6}$$

Similarly, the theoretical pump flow rate in terms of a litre per minute can be expressed using the factor 1 L equal to 1000 cm³ as,

$$Q_T = \frac{V_P \times N}{1000} \text{ lpm} \tag{4.7}$$

**Example 4.2** The displacement of a pump is 3 in.³ per revolution. Find the theoretical flow rate if the pump is driven at 1800 rpm.

**Solution**
The pump flow rate is calculated as,

$$Q_T = V_P \times N = 3 \times 1800 = 5400 \text{ in.}^3/\text{min} = \frac{5400}{231} = 23.38 \text{ gpm} \qquad (4.8)$$

**Practice Problem 4.2**
A pump is driven by a motor at a speed of 1200 rpm and the flow rate is 10 gpm. Calculate the pump displacement.

## 4.4 Pump Power and Torque

An induction motor converts electrical power to mechanical power to drive or rotate the pump. This rotational speed reduces by the pressure at the delivery pipe. Therefore, the induction motor should have enough power to overcome this problem. The torque is produced when the pump-motor is rotating in a specified direction as shown in Fig. 4.4.

The circumference distance is expressed as,

$$d = 2\pi r \text{ (ft or m)} \qquad (4.9)$$

The energy is expressed as,

$$\text{Energy} = F \times d \text{ (ft-lb or N-m or J)} \qquad (4.10)$$

Substituting Eq. (4.9) into Eq. (4.10) yields,

$$\text{Energy} = F \times 2\pi r \text{ (ft-lb or N-m or J)} \qquad (4.11)$$

The torque that is generated at the shaft of the motor pump is defined as the product of force and distance. Mathematically, it is written as,

$$T = F \times r \text{ (ft-lb or N-m)} \qquad (4.12)$$

Substituting Eq. (4.12) into Eq. (4.11) yields,

**Fig. 4.4** A pump shaft with a force

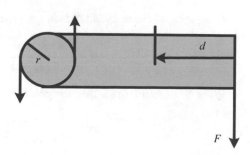

$$\text{Energy} = 2\pi T \ (\text{ft-lb or N-m or J}) \tag{4.13}$$

The rate of receiving and delivering energy is known as power ($p$) and mathematically, it is expressed as,

$$p = \frac{dW}{dt} \tag{4.14}$$

In general, the power is expressed as,

$$p = \frac{W}{t} = \frac{\text{Energy}}{\text{time}} \ (\text{J/s}) \tag{4.15}$$

Again, consider $N$ is the rotation speed of the pump in revolution per minute (rpm). Equation (4.15) can be rearranged as [3],

$$p = \frac{\text{Energy}}{\text{rev}} \times \frac{\text{rev}}{\text{time}} \tag{4.16}$$

Substituting Eq. (4.13) into Eq. (4.16) yields,

$$p = 2\pi T \times N \tag{4.17}$$

In Eq. (4.17), the unit of speed is rpm. Therefore, it needs to convert rpm to rps to get the unit of power in ft-lb/s or N-m/s or J/s. Equation (4.17) can be written as,

$$p = 2\pi T \times N \times \frac{1 \text{ min}}{60 \text{ s}} \tag{4.18}$$

$$p = \frac{T \times N}{9.55} \tag{4.19}$$

If the torque is in ft lb and $N$ is in rpm, then Eq. (4.19) is modified by using the conversion factor 550 ft-lb/s equals to one HP as,

$$HP = \frac{T \times N}{9.55 \times 550} = \frac{T \times N}{5252} \tag{4.20}$$

If the torque is in in lb and $N$ is in rpm, then Eq. (4.20) is modified as,

$$HP = \frac{T \times N}{5252 \times 12} = \frac{T \times N}{63024} \tag{4.21}$$

When SI units are used, then the power is calculated as,

$$p(\text{kW}) = \frac{T \times N}{9.55 \times 1000} = \frac{T \times N}{9550} \tag{4.22}$$

The pump torque is fully dependent on the system pressure and pumps displacement. Mathematically, it is expressed as,

$$T = \frac{P \times V_p}{2\pi} \tag{4.23}$$

where,

$P$ is the system pressure in units of psi,
$V_p$ is the pump displacement in units of in.$^3$/rev,
$T$ is the torque in units of lb. in.

***Example 4.3*** A hydraulic pump with a displacement of 2 in.$^3$ per revolution is selected for a 1500 psi system. Calculate the value of the driving torque if the efficiency is 100%.

**Solution**
The value of the driving torque is calculated as,

$$T_d = \frac{P \times V_P}{2\pi} = \frac{1500 \times 2}{2\pi} = 477.46 \, \text{in. lb} \tag{4.24}$$

**Practice Problem 4.3**
A 500 N m is required to operate a system whose pressure is 125 Pa. Determine the pump displacement.

## 4.5 Pumps Classification

A hydraulic pump is used to pump when a certain volume of fluid is displaced and transferred from one place to another place. According to displacement, pumps are classified as positive displacement (hydrostatic) and non-positive (hydrodynamic or kinetic or dynamic pressure. When the pumping action displaces a specified amount of fluid per revolution is known as a positive displacement pump. There will be a definite amount of fluid discharge irrespective of an increase in pressure by a moving member such as a piston. Hence, only PDP can be used for the hydraulic system. The positive displacement pump is classified as a reciprocating and rotary pump [4].

The reciprocating pumps are classified based on the action of water on the piston and the number of cylinders uses.

Based on the action of water on the piston

- Single-acting pump
- Double-acting pump.

Based on the number of cylinders uses

- Single-cylinder pump
- Double cylinder pump
- Triple cylinder pump.

Rotary pumps are classified as

- Gear pump
- Vane pump
- Lobe pump
- Piston pump
- Screw pump.

Pumps that are using the inertia principle to push the fluid are known as nonpositive displacement pumps or dynamic pressure pumps. The discharge reduces with an increase in pressure and the discharge becomes zero at the maximum pressure. Therefore, NPDP is not used for the hydraulic system. This type of pump is generally used for low-pressure (250–300 psi), high-volume flow applications. Dynamic pressure pumps are again classified as centrifugal, turbine, axial flow propeller pumps.

## 4.6  Working Principle of Pump

Initially, the motor converted electrical energy into mechanical energy. This mechanical action of a pump increases the volume which in turn creates a low-pressure vacuum at the inlet which allows the atmospheric pressure to force liquid from the main reservoir into the inlet line to the pump as shown in Fig. 4.5. It is also known as the suction side of the pump and the pressure of the liquid is lowest at the suction side.

Finally, the same mechanical action of a pump delivers this liquid to the pump outlet and forces it into the hydraulic system. The liquid leaves the pump through the outlet and this side is called the discharge side. The pressure of the liquid is highest at the discharge side.

**Fig. 4.5**  A simple pump

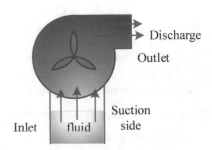

## 4.7 Rotary Pumps

All rotary pumps have rotating parts, which trap the fluid at the inlet port and force it, through the discharge port, into the fluid power systems. Gears, screws, lobes, and vanes are commonly used to move the fluid within the pump.

Rotary pumps are designed with very small clearances between their rotating and stationary parts to minimize slippage from the discharge side back to the suction side of the pump. They are designed to operate at relatively moderate speeds, normally below 1800 rpm. Operation at higher speeds can cause erosion and excessive wear. There are numerous types of rotary pumps and various methods of classification. They may be classified by the shaft position, the type of driver, their manufacturer's name, or their service application. However, the classification of rotary pumps is generally made according to the type of rotating element.

### 4.7.1 Gear Pump

A gear pump develops flow by carrying fluid between the teeth of two meshed gears as shown in Fig. 4.6. One gear is driven by the driveshaft and turns the other. The pumping chambers formed between the gear teeth are enclosed by the pump's housing and the side plates. A partial vacuum is created at the inlet as the gear teeth unmeshed. Fluid flows in to fill the space and is carried around the outside of the gears. As the teeth mesh again at the outlet, the fluid is forced out. The high pressure at the pump's outlet might impose an unbalanced load on the gears and their associated bearing support structure.

The flow rate of this pump is calculated by considering the pump as a hollow cylinder of outside and inside diameters and the width of the gear teeth. According

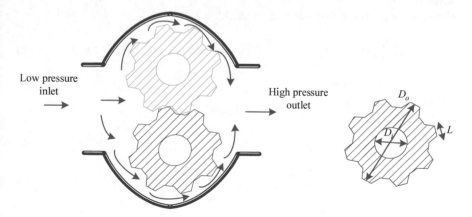

**Fig. 4.6** A simple gear pump

to Fig. 4.6, consider $D_o$ is the outside diameter of gear teeth (in., m), $D_i$ is the inside diameter of gear teeth (in., m), $L$ is the width of gear teeth (in., m), $V_d$ is the displacement volume of the pump (in³/rev, m³/rev), N is the rpm of the pump. The volumetric displacement is calculated as,

$$V_d = \frac{\pi}{4}\left(D_o^2 - D_i^2\right) \times L \tag{4.25}$$

According to Eq. (4.6), the theoretical pump flow rate in in.³/min and m³/min can be expressed as,

$$Q_T = V_d \times N = \frac{\pi}{4}\left(D_o^2 - D_i^2\right) \times L \times N \tag{4.26}$$

The value of the pump flow rate in gpm is calculated (1 gallon = 231 in.³) as,

$$Q_T = \frac{\frac{\pi}{4}\left(D_o^2 - D_i^2\right) \times L \times N}{231} \text{ gpm} \tag{4.27}$$

From Eq. (4.26), it is seen that the pump flow rate directly varies with the pump speed as explained in Fig. 4.7.

At a given speed, the pump theoretical flow rate is constant with respect to system pressure that is explained in Fig. 4.8. Whereas the actual flow rate is less than the theoretical flow rate due to frictional losses or internal losses.

**Example 4.4** The inside and outside diameters of a gear pump are 1.5 in. and 2.5 in., respectively. The pump speed and gear width are found to be 1800 rpm, and 1 in., respectively. Calculate the pump flow rate.

**Solution**
The value of the pump displacement is calculated as,

$$V_d = \frac{\pi}{4}\left(D_o^2 - D_i^2\right) \times L = \frac{\pi}{4}\left(2.5^2 - 1.5^2\right) \times 1 = 3.14 \text{ in.}^3/\text{rev} \tag{4.28}$$

**Fig. 4.7** Flow rate versus pump speed

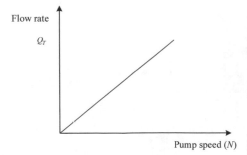

**Fig. 4.8** Flow rate with pressure and a given pump speed

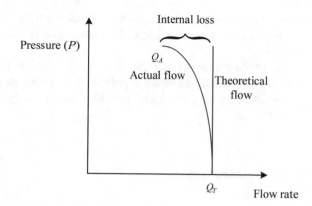

The pump flow rate is calculated as,

$$Q_T = \frac{N \times V_P}{231} = \frac{1800 \times 3.14}{231} = 24.47 \text{ gpm} \qquad (4.29)$$

**Practice Problem 4.4**

The inside and outside diameters of a gear pump are 1.2 in. and 2.6 in., respectively. Calculate the pump speed if the gear width is found to be 0.95 in. Consider the flow rate of 20 gpm.

### 4.7.2 Lobe Pump

The lobe pumps are considerably larger than gear teeth pumps, but there are only two or three lobes are used on each rotor. A three-lobed pump is illustrated in Fig. 4.9. The two elements of the lobe pump are rotated, one is directly driven by the power source, and the other is driven by timing gears. As the elements rotate, the liquid is

**Fig. 4.9** A simple lobe pump

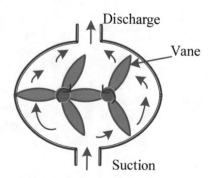

trapped between two lobes of each rotor and the walls of the pump chamber. Then, the trapped liquid is carried from the suction side to the discharge side of the pump chamber.

As liquid leaves, the suction chamber, the pressure in the suction chamber is lowered and additional liquid is pulled into the chamber from the reservoir. Lobe pumps are used in a variety of industries such as pulp and paper, chemical, food, beverage, etc.

### 4.7.3   Vane Pump

The simplest vane pump consists of a circular rotor rotating inside of a larger circular cavity. The centers of the two circles are balanced, causing eccentricity as shown in Fig. 4.10. The leakage in a gear pump creates from the small gaps between the teeth as well as between teeth and pump housing. The vane pump reduces this leakage by using spring-loaded vanes slotted into a driven rotor.

The vanes of the pump can slide into and out of the rotor and seal on all edges. As a result, creating vane chambers for doing the pumping work. In the first half of the revolution (intake side of the pump), the volume between the rotor and the cam ring increases. These increasing volume vane chambers are filled with fluid forced in by the inlet pressure. This inlet pressure is the pressure from the system being pumped, often refer just to the atmosphere. As the rotation continues, the volume of the cavity is reduced. In the next half of the revolution, when the pumping chamber reaches the outlet port, the fluid is pushed out of the pump and into the system.

The relevant parameters of a vane pump are shown in Fig. 4.11. Consider the following parameters to derive the flow rate of a vane pump.

$D_r$ is the diameter of the rotor in m,
$D_c$ is the diameter of the cam ring in m.
$e$ is the eccentricity in m,
$L$ is the width of the rotor in mm.

From Fig. 4.11, the maximum possible eccentricity is expressed as,

**Fig. 4.10**  A simple vane pump

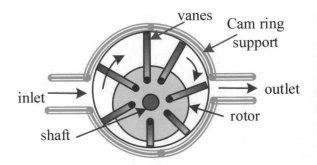

**Fig. 4.11** Different
diameters of a vane pump

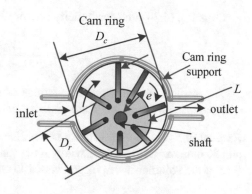

$$e_{max} = \frac{D_c - D_r}{2} \tag{4.30}$$

This maximum value of eccentricity produces the maximum value of pump volumetric displacement and it can be expressed as,

$$V_{Dm} = \frac{\pi}{4}(D_c^2 - D_r^2) \times L \tag{4.31}$$

$$V_{Dm} = \frac{\pi}{4}(D_c + D_r)(D_c - D_r) \times L \tag{4.32}$$

Substituting Eq. (4.29) into Eq. (4.32) yields,

$$V_{Dm} = \frac{\pi}{4}(D_c + D_r)(2e_m) \times L \tag{4.33}$$

$$V_{Dm} = \frac{\pi}{2}(D_c + D_r)(e_m) \times L \tag{4.34}$$

The actual volumetric displacement occurs when $e_{max} = e$ and Eq. (4.34) is gaian modified as,

$$V_{Dm} = \frac{\pi}{2}(D_c + D_r)(e) \times L \tag{4.35}$$

Then the theoretical discharge is expressed as,

$$Q_T = \frac{\pi}{2}(D_c + D_r)(e)LN \tag{4.36}$$

The eccentricity i.e. the deviation from the normal pattern is expressed as,

$$e = \frac{2Q_T}{\pi(D_c + D_r)LN} \tag{4.37}$$

From Eq. (4.34), the eccentricity is expressed as,

$$e = \frac{2V_{dm}}{\pi(D_c + D_r)L} \tag{4.38}$$

The displacement and pressure ratings of a vane pump are generally lower than gear pumps. However, its volumetric efficiency is around 95%.

***Example 4.5*** The cam ring and the rotor diameters of a vane pump are found 60 mm and 80 mm, respectively. The width of the vane is 40 mm. Calculate the eccentricity for it if the volumetric displacement of 100 cm$^3$.

**Solution**
The value of the eccentricity is calculated as,

$$e = \frac{2V_{dm}}{\pi(D_c + D_r)L} = \frac{2 \times 100}{\pi(6 + 8) \times 4} = 1.14\,\text{cm} \tag{4.39}$$

**Practice Problem 4.5**
The cam ring and the rotor diameters of a vane pump are found 50 mm and 70 mm, respectively. Calculate the width of the vane if the eccentricity and the volumetric displacement are found to be 2.5 cm and 155 cm$^3$, respectively.

### 4.7.4 Screw Pump

The screw pump is an axial flow positive displacement unit. The two shafts namely the driving shaft and driven shaft are attached to form meshing within a close-fitting housing as shown in Fig. 4.12 The blades are attached to the shafts.

The driving and driven shafts are connected by a timing gear. The driving shaft is coupled with a motor. The two screws are forming closed chambers that are moving

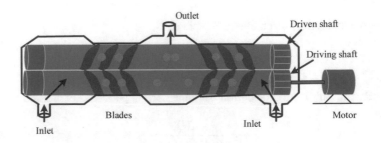

**Fig. 4.12** A simple screw pump

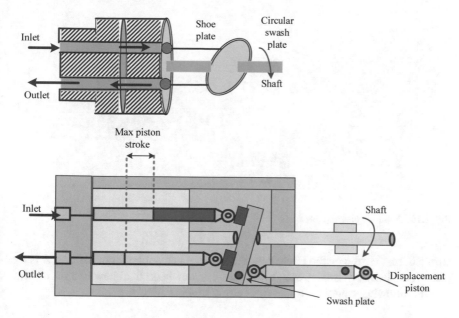

**Fig. 4.13**  A simple piston pump

in an axial direction during turning. This movement creates a vacuum at the inlet side and pressure at the outlet. The water through the inlets due to atmospheric pressure and delivers through the discharge pipe.

### 4.7.5  Piston Pump

In a piston pump, the cylinder block and driveshaft are located on the same centerline as shown in Fig. 4.13. The piston usually moves backward direction to create the vacuum at the inlet pipe. This vacuum pressure allows atmospheric pressure to push fluid from the tank or reservoir into the pump. After that, the piston moves forward to expel fluid from the discharge pipe. The check valves are used in both the suction and delivery pipes to control the fluid. This type of pump is normally designed with a variable displacement capability. The maximum swash plate angle is limited to 17.5° by construction.

## 4.8  Theoretical Discharge of Axial Piston Pump

The inlet and outlet positions of pistons of an axial piston pump are shown in Fig. 4.14. The piston stroke varies with respect to the offset angle, $\theta$. Consider $S$ is the piston

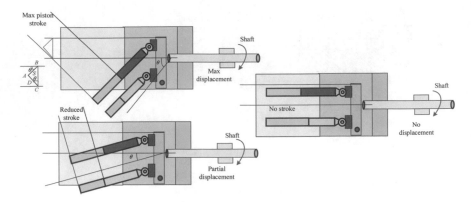

**Fig. 4.14** A simple piston pump

stroke in m, $D$ is the piston circle diameter in m, $n$ is the number of pistons, $A$ is the piston area in m$^2$, $N$ is the piston rpm and $Q_T$ is the theoretical flow rate in m$^3$/min.

The following equation from a right angle triangle ABC is expressed as,

$$\tan \theta = \frac{AB}{AC} \tag{4.40}$$

$$\tan \theta = \frac{S}{D} \tag{4.41}$$

$$S = D \times \tan \theta \tag{4.42}$$

The displacement volume of one piston is,

$$V_d = A \times S \, \text{m}^3 \tag{4.43}$$

The total displacement of $n$ number of pistons is expressed as,

$$V_d = n \times A \times S \, \text{m}^3/\text{rev} \tag{4.44}$$

Substituting Eq. (4.42) into Eq. (4.44) yields,

$$V_d = n \times A \times D \times \tan \theta \, \text{m}^3/\text{rev} \tag{4.45}$$

Then the theoretical flow rate is expressed as,

$$Q_T = n \times A \times D \times \tan \theta \times N \, \text{m}^3/\text{min} \tag{4.46}$$

***Example 4.6*** An eight-bore cylinder of a fixed displacement axial piston pump operates at 1800 rpm. The diameter of each bore is 12 mm and the stroke length is 16 mm. Calculate the theoretical flow rate.

**Solution**
The value of the theoretical displacement is calculated as,

$$Q_T = \pi \left(\frac{D^2}{4}\right) \times LNn = \pi \left(\frac{0.012^2}{4}\right) \times 0.016 \times 1800 \times 8 = 0.03 \text{ m}^3/\text{min}$$

(4.47)

**Practice Problem 4.6**
A nine-bore cylinder variable displacement axial piston pump operates at 1800 rpm. The diameter of each bore is 10 mm and the stroke length is 15 mm. Calculate the theoretical flow rate if the offset angle 12°.

## 4.9 Centrifugal Pump

A simple schematic diagram of a centrifugal pump is shown in Fig. 4.15. The impeller, impeller vane, eye, shaft, casing, inlet pipe, outlet pipe, discharge valve, seat valve and strainer or filter are mentioned here. The function of the strainer allows fluid into the suction pipe without foreign materials. The seat valve allows fluid into the suction pipe. The static head of the pump is equal to the sum of the suction head and the discharge head.

Fluid enters the pump through the suction pipe and moves around the pump casing which creates the tangential velocity. The magnitude of this tangential velocity is low around the casing as the area is higher. Whereas this velocity is lower in the discharge pipe as the area is higher which satisfies the continuity equation ($Q = AV = $ constant).

Bernoulli's equation in energy form is used to explain the working principle and this equation is written as,

$$\text{Pressure energy} + \text{kinetic energy} \left(\frac{mv^2}{2}\right) + \text{potential head energy} = \text{constant}$$

(4.48)

$$gh \text{ (potential energy)} + \frac{P_1}{\rho} \text{ (pressure energy)} + \frac{1}{2}v^2 \text{ (kinetic energy)} = \text{constant}$$

(4.49)

The potential head energy is zero as the suction head constant with respect to the datum line. Therefore, Eq. (4.49) is revised as,

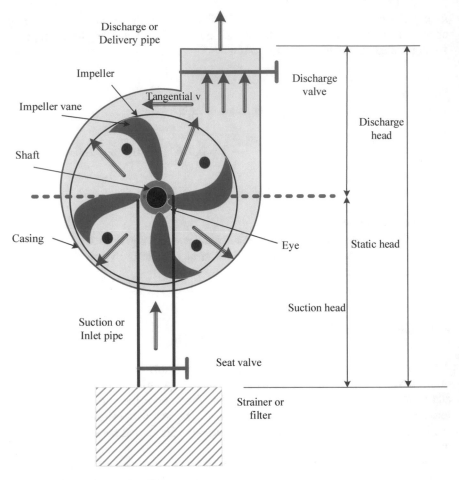

**Fig. 4.15** A simple centrifugal pump

$$\text{Pressure energy} + \text{kinetic energy} = \text{constant} \qquad (4.50)$$

In the discharge pipe, the velocity is low so that the kinetic energy is low. According to Eq. (4.50), pressure energy is high. As a result, the fluid discharged through the delivery pipe. Therefore, the fluid will discharge from the outlet with high pressure.

## 4.10   Velocity Triangle of Centrifugal Pump

A velocity triangle for a centrifugal pump is considered to find work done by the impeller. This velocity triangle with inlet and outlet terminals of a vane of an impeller is shown in Fig. 4.16. Assume that there is a steady fluid flow, no energy loss due to friction and shock at the entry point of the impeller.

The following parameters are defined as,

$D_i$ is the diameter of the impeller at the inlet in m,
$D_o$ is the diameter of the impeller at the outlet in m,
$N$ is the speed of the impeller in rpm,
$u_i$ is the tangential velocity of the impeller at the inlet in m/s,
$V_i$ is the absolute velocity of the liquid at the inlet in m/s,
$V_o$ is the absolute velocity of the liquid at the inlet in m/s,
$V_{ri}$ is the relative of the liquid at the inlet in m/s,
$V_{ro}$ is the relative of the liquid at the outlet in m/s,
$V_{wi}$ and $V_{wo}$ are velocities of the whirl (rotation) at the inlet and outlet in m/s,

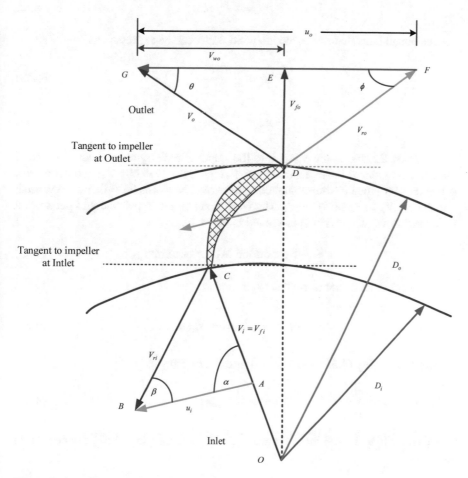

**Fig. 4.16** A velocity triangle of a centrifugal pump

$\alpha$ is the angle made by absolute velocity $V_i$ at the inlet with the direction of motion of the vane in deg,

$\theta$ is the angle made by absolute velocity $V_o$ at the outlet with the direction of motion of the vane in deg,

$\beta$ is the angle made by relative velocity $V_{ri}$ at the inlet with the direction of motion of the vane in deg,

$\beta$ and $\phi$ are the vane angles at the inlet and the outlet in deg,

$B_i$ and $B_o$ are the widths of the impeller at the inlet and outlet in m.

The areas of the impeller at the inlet and outlet are expressed as,

$$A_i = \pi D_i B_i \tag{4.51}$$

$$A_o = \pi D_o B_o \tag{4.52}$$

The tangential velocities at the inlet and outlet can be expressed as,

$$u_i = \frac{\pi D_i N}{60} \tag{4.53}$$

$$u_o = \frac{\pi D_o N}{60} \tag{4.54}$$

Assumed that the water enters into the impeller radially at the inlet. Therefore, the whirl component $V_{wi}$ equals zero and $V_i$ equals flow velocity $V_{fi}$. In other words, if $\alpha$ equals 90°, then the horizontal component $V_{wi} = V_i \cos 90 = 0$ and the vertical component $V_{fi} = V_i \sin 90 = V_i$. The work done by the impeller liquid per second unit weight of water striking per second is given by,

$$w = -(\text{work done in case of turbine}) \tag{4.55}$$

The work done by the turbine is expressed as,

$$w_T = \frac{1}{g}(V_{wi}u_i - V_{wo}u_o) \tag{4.56}$$

Substituting Eq. (4.56) and $V_{wi} = 0$ into Eq. (4.55) yields,

$$w = \frac{1}{g}V_{wo}u_o \tag{4.57}$$

Consider $W$ is the weight of the water per second, then Eq. (4.57) can be revised as,

$$w = \frac{W}{g}V_{wo}u_o \tag{4.58}$$

Again substituting $W = \rho_w g Q$ in Eq. (4.58) yields,

$$w = \rho_w Q V_{wo} u_o \qquad (4.59)$$

where $Q$ ia the volume of water per second (flow rate).

The flow rate is expressed as,

$$Q = \pi D_i B_i V_{fi} = \pi D_o B_o V_{io} \qquad (4.60)$$

The torque is the rate of change of angular momentum and it is expressed as,

$$T = \rho_w Q V_{wo} R_o \qquad (4.61)$$

where $R_o$ is the outlet radius of the impeller.

Therefore, the power at the impeller is expressed as,

$$P_{imp} = \omega T \qquad (4.62)$$

Substituting Eq. (4.61) into Eq. (4.61) yields,

$$P_{imp} = \rho_w Q V_{wo} \omega R_o \qquad (4.63)$$

Again, substituting $u_o = R_o \omega$ into Eq. (4.63) yields,

$$P_{imp} = \rho_w Q V_{wo} u_o \qquad (4.64)$$

The unit of impeller power is W (watt) and then it is expressed in kW as,

$$P_{imp} = \frac{\rho_w Q V_{wo} u_o}{1000} \text{ kW} \qquad (4.65)$$

From Eq. (4.65), the impeller power can be calculated if other parameters are given.

## 4.11 Graphic Symbols of Pump

The function of a hydraulic pump is to transform mechanical energy into hydraulic energy. Different types of pump symbols are used in hydraulic systems. These are fixed displace displacement, variable displacement, pressure compensated. The displacement of a pump is defined by the volume that the pump gears, vanes or pistons will displace in one revolution. The pump that has a fixed displacement per revolution is known as a fixed displacement pump or unidirectional pump. The unidirectional pump normally rotates in one direction. If rotating these motors in the wrong direction may result in a blown shaft seal and the motor drainage the fluid through the suction (inlet) port. In any case, blocking the outlet of a fixed displacement pump will send high flow to the tank through the pressure relief valve. The small solid triangle represents the direction of flow as shown in Fig. 4.17.

A pump that can increase or decrease the displacement either manually, hydraulically or electronically is known as a variable displacement pump as shown in Fig. 4.18. The flow rate can vary of a variable displacement pump at a constant speed. The pump that automatically reduces their displacement to prevent the pressure from rising from the preset level is known as the pressure compensated variable displacement pump.

The arrow that is parallel to the flow path indicates the pressure compensated as shown in Fig. 4.19.

The flow rate of a pressure compensated pump automatically decreases when pressure increases. This type of pump is only available in vane and piston designs.

**Fig. 4.17**  A symbol of a
fixed displacement pump

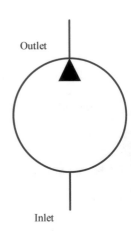

**Fig. 4.18**  A symbol of a
variable displacement pump

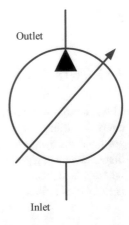

**Fig. 4.19** A symbol of a pressure compensated variable displacement pump

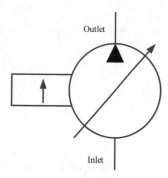

**Fig. 4.20** A symbol of a bi-directional displacement pump

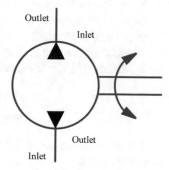

The pump that has two energy triangles to show that the fluid is coming out from both ports is known as a bi-directional displacement pump. At a time, only one output is taken from one port while the opposite port is considered as an inlet as shown in Fig. 4.20. Variable flow rate can be obtained at a constant speed and this pump can rotate in either direction. Based on the requirement, this type of pump is used in a hydraulic system.

## 4.12 Performance of Pump

The performance of any device is identified by its efficiency. This efficiency depends on the input power and output power. However, the output power depends on the system loss. In the case of the pump, there are some components inside the pump casing. The losses in the pump casing should be minimum. There are three types of efficiency are considered in a pump. These are volumetric, mechanical and overall efficiencies.

The volumetric efficiency refers to the amount of leakage that occurs during the operation of a pump. This volumetric efficiency is defined as the ratio of the actual flow rate $(Q_A)$ to the theoretical flow rate $(Q_T)$ of a pump. Mathematically, it is expressed as,

**Table. 4.1** Volumetric efficiency of different types of pump

| Types of pump | Volumetric efficiency (%) |
| --- | --- |
| Gear pumps | 80–90 |
| Vane pumps | 92 |
| Piston pumps | 90–98 |

$$\eta_v = \frac{Q_A}{Q_T} \times 100 \qquad (4.66)$$

The values of the volumetric efficiency for different types of pumps are mentioned in Table 4.1.

Mechanical efficiency is defined as the ratio of actual pump power output to the input power of the shaft. This actual pump power output is always less than the shaft input power due to friction in bearings, mating parts and shocks or vibration when water entering the pump. The mechanical efficiency of a pump varies from 90 to 95%. Mathematically, the mechanical efficiency is expressed as,

$$\eta_m = \frac{\text{Power delivered to the fluid}}{\text{Power required to drive the pump}} = \frac{P_w}{P_s} \qquad (4.67)$$

The power delivered to the fluid is traditionally known as water horsepower and it is expressed as,

$$P_w = p \, Q_T \qquad (4.68)$$

where $p$ is the pump discharge pressure in Pa (N/m$^2$) and $Q_T$ is the theoretical flow rate in m$^3$/s.

The power required to drive the pump is known as shaft power or brake horsepower and it can be expressed as,

$$P_s = \omega \, T \qquad (4.69)$$

where $\omega$ is the pump shaft angular speed rad/s, $T$ is the pump shaft torque in Nm. Again, consider $T_A$ is the actual torque delivered to the pump in Nm and $\omega = N$ is the speed of the pump in rad/s. Then Eq. (4.69) is revised as,

$$P_s = T_A N \qquad (4.70)$$

Substituting Eqs. (4.69) and (4.70) into Eq. (4.67) yields,

$$\eta_m = \frac{p \, Q_T}{T_A N} \qquad (4.71)$$

The unit in rad per second (rad/s) is converted in the following ways,

$$\text{rad/s} \times \text{s/min} \times \text{rev/rad} = \text{rpm} \tag{4.72}$$

The radian and revolution is related as,

$$2\pi \text{ rad} = 1 \text{ rev} \tag{4.73}$$

Substituting Eq. (4.73) and 1 min = 60 s in Eq. (4.73) yields,

$$N \text{ rad/s} \times \text{s/60 s} \times \frac{1}{2\pi} \text{ rev/rad} = N \text{ rpm} \tag{4.74}$$

$$N \text{ rad/s} = \frac{2\pi}{60} N \text{ rpm} \tag{4.75}$$

The input power delivered to the pump shaft is equal to the product of the torque $(T_A)$ and the rotational speed. Mathematically, it is expressed as,

$$P_s = T_A(\text{ft-lb}) \times N(\text{rad/s}) = \frac{2\pi}{60} T_A \times N \text{ (rpm)} = \frac{T_A \times N}{60/2\pi} = \frac{T_A \times N}{9.549} \text{ W} \tag{4.76}$$

Using the conversion factor 1 HP = 550 ft-lb/s = 746 N-m/s (W) to modify Eq. (4.76),

$$P_s = \frac{T_A \times N}{9.545 \times 550} \text{ HP} \tag{4.77}$$

$$P_s = \frac{T_A \times N}{5252} \text{ HP} \tag{4.78}$$

where the torque $(T_A)$ is in ft-lb and it is converted to in.-lb as shown in Eq. (4.78),

$$P_s = \frac{T_A \times N}{5252 \times 12} \text{ HP} \tag{4.79}$$

$$P_s = \frac{T_A \times N}{63025} \text{ HP} \tag{4.80}$$

Again, the mechanical efficiency is expressed as the ratio of theoretical torque required to operate the pump to the actual torque delivered to the pump and it is written as,

$$\eta_m = \frac{\text{Theoretical toeque required to operate pump}}{\text{Actual torque delivered to pump}} = \frac{T_T}{T_A} \tag{4.81}$$

According to Eq. (4.23), the theoretical torque is determined as,

$$T_T = \frac{p(\text{psi}) \times V_p(\text{in.}^3)}{2\pi} \ (\text{lb-in.}) \tag{4.82}$$

$$T_T = \frac{p(\text{Pa}) \times V_p(\text{m}^3)}{2\pi} \ (\text{N-m}) \tag{4.83}$$

The actual torque is determined as,

$$T_A = \frac{\text{power delivered to pump}(p)}{\omega} = \frac{p \ (\text{N-m/s})}{\omega \ (\text{rad/s})} \ (\text{N-m}) \tag{4.84}$$

where,

$$\omega \ \text{rad/s} = \frac{2\pi}{60} \ N \ \text{rpm} \tag{4.85}$$

where the speed $N$ is in rpm.

The overall efficiency is defined as the ratio of actual power delivered by the pump to the actual power delivered to the pump shaft and mathematically, it is written as,

$$\eta_m = \frac{\text{Theoretical toeque required to operate pump}}{\text{Actual torque delivered to pump}} = \frac{T_T}{T_A} \tag{4.86}$$

The overall efficiency is also defined as the product of volumetric efficiency and mechanical efficiency and mathematically, it is expressed as,

$$\eta_o = \eta_v \times \eta_m \tag{4.87}$$

Substituting Eqs. (4.66) and (4.71) into Eq. (4.87) yields,

$$\eta_o = \frac{Q_A}{Q_T} \times \frac{p \ Q_T}{T_A N} \tag{4.88}$$

$$\eta_o = \frac{p \ Q_A}{T_A N} \tag{4.89}$$

$$\eta_o = \frac{\text{Actual power delivered by the pump}}{\text{Actual power delivered to pump}} = \frac{p \ Q_A}{T_A N} \tag{4.90}$$

According to US Customary, the actual power delivered to the pump from the three-phase induction motor is known as input horsepower or brake horsepower. In general, it is expressed as,

$$P_{in} = \frac{T \times N}{63,025} \ \text{HP} \tag{4.91}$$

where $T$ is in.-lb and $N$ is in rpm.

In the SI unit, from Eq. (4.76), the expression of input power is expressed as,

$$P_{in} = \frac{T \times N}{9.55} W = \frac{T \times N}{9.55 \times 1000} kW = \frac{T \times N}{9550} kW \qquad (4.92)$$

where $T$ is N-m and $N$ is in rpm.

According to US Customary, the actual power delivered to the hydraulic fluid is known as hydraulic horsepower. In general, it is expressed as,

$$P_{hyd} = \frac{p \times Q}{1714} HP \qquad (4.93)$$

where $p$ is in psi and $Q$ is in gpm.

In the SI unit, it is expressed as,

$$P_{hyd} = \frac{p \times Q}{60,000} kW \qquad (4.94)$$

where $p$ is kPa and $Q$ is in lpm.

***Example 4.7*** A 6 in.$^3$ displacement pump delivers 25 gpm at 1200 rpm and 900 psi. Calculate the theoretical torque required to operate the pump and the overall efficiency if the input torque of the pump is 800 in.-lb.

**Solution**

The value of the theoretical flow rate is calculated as,

$$Q_T = \frac{V_d \times N}{231} = \frac{6 \times 1200}{231} = 31.17 \, gpm \qquad (4.95)$$

The volumetric efficiency is calculated as,

$$\eta_v = \frac{Q_A}{Q_T} = \frac{25}{31.17} = 0.80 \qquad (4.96)$$

The value of the theoretical torque is calculated as,

$$\eta_v = \frac{T_T}{T_A} = \frac{T_T}{800} \qquad (4.97)$$

$$T_T = \eta_v \times 800 = 0.8 \times 800 = 640 \, in.\text{-}lb \qquad (4.98)$$

The input power of the pump is calculated as,

$$P_{in} = \frac{T_A \times N}{63024} = \frac{800 \times 1200}{63,024} = 15.23 \, HP \qquad (4.99)$$

The output power by the pump is calculated as,

$$P_{hyd} = \frac{p \times Q_T}{1714} = \frac{900 \times 31.17}{1714} = 16.37\,\text{HP} \tag{4.100}$$

The mechanical efficiency is calculated as,

$$\eta_m = \frac{P_{hyd}}{P_{in}} = \frac{15.23}{16.37} = 0.93 \tag{4.101}$$

The overall efficiency is calculated as,

$$\eta_o = \eta_v \times \eta_m = 0.8 \times 0.93 = 0.74 \tag{4.102}$$

## Practice Problem 4.7

A pump is driven by a 14 HP prime and delivers fluid at 30 lpm at a pressure of 9 MPa. Find the overall efficiency.

## Exercise Problems

4.1    The number of poles of a three-phase induction motor is 2. Find the speed of the pump if the frequency is 60 Hz.

4.2    The displacement of a pump is 4 in.$^3$ per revolution. Find the theoretical flow rate if the pump is driven at 1200 rpm.

4.3    A pump is driven by a motor at a speed of 1800 rpm and the flow rate is 15 gpm. Determine the pump displacement.

4.4    A hydraulic pump with a displacement of 4 in.$^3$ per revolution is selected for a 1200 psi system. Calculate the value of the driving torque if the efficiency is 100%.

4.5    A 800 N m is required to operate a system whose pressure is 145 Pa. Calculate the pump displacement.

4.6    The inside and outside diameters of a gear pump are 2.5 in. and 3.5 in., respectively. The pump speed and gear width are found to be 1200 rpm, and 1 in., respectively. Calculate the pump flow rate.

4.7    The inside and outside diameters of a gear pump are 1 in. and 2.5 in., respectively. Calculate the pump speed if the gear width is found to be 0.90 in. Consider the flow rate of 10 gpm.

4.8    The cam ring and the rotor diameters of a vane pump are found 50 mm and 70 mm, respectively. The width of the vane is 30 mm. Calculate the eccentricity for it if the volumetric displacement of 100 cm$^{3.}$

4.9    The cam ring and the rotor diameters of a vane pump are found 25 mm and 50 mm, respectively. Calculate the width of the vane if the eccentricity and the volumetric displacement are found to be 3.5 cm and 125 cm$^3$, respectively.

4.10 An six-bore cylinder of a fixed displacement axial piston pump operates at 1200 rpm. The diameter of each bore is 10 mm and the stroke length is 16 mm. Calculate the theoretical flow rate.

4.11 A four-bore cylinder variable displacement axial piston pump operates at 1800 rpm. The diameter of each bore is 12 mm and the stroke length is 15 mm. Calculate the theoretical flow rate if the offset angle is 15°.

4.12 Compute the theoretical speed of a fluid power motor with a 2.5 in.$^3$ per revolution displacement receiving fluid at the rate of 8.5 GPM.

# References

1. F. Don Norvelle, *Fluid Power Technology* (West Publishing Company, New York, USA, 1995), pp. 1–649
2. M.G. Rabie, *Fluid Power Engineering* (McGraw-Hill, New York, USA, 2009), pp. 1–420
3. A. Parr, *Hydraulics and Pneumatics—A Technician's and Engineer's Guide*, 3rd edn. (Elsevier, New York, USA, 2011)
4. J.A. Sullivan, *Fluid Power Theory and Applications*, 3rd edn. (Prentice-Hall Inc., New Jersey, USA, 1989), pp. 1–528

# Chapter 5
# Hydraulic Directional Valves

## 5.1 Introduction

The control of fluid flow in hydraulic and pneumatic circuits is an important consideration to improve efficiency and reduce losses. This important consideration is to select the components such as the valve. There are three basic types of valves are used in the fluid power system. These are directional control valves (DCV), pressure control valves (PCV), and flow control valves (FCV). The directional control valves control the direction of the fluid. The check valve, shuttle valves, two-way, three-way, and four-way directional control valves are used. The pressure control valves are used in a fluid power system to protect different types of components against overpressure due to the sudden closing of the valves. The pressure control in a fluid power system is accomplished by pressure reducing, pressure relief, sequence, unloading, and counterbalance valves. The flow control valve controls the amount of fluid flow (flow rates) in hydraulic and pneumatic circuits. The speed of the cylinder mainly depends on the amount of fluid that passes through the tubing or piping. This type of task is accomplished by the flow control valve such as a one-way control valve or non-return throttle valve. This chapter will discuss different types of control valves, classification, construction, and applications. In addition, Automation Studio software is used to simulate the operation of cylinders with the directional control valves.

## 5.2 Control Valves

The control part is the most important characteristic of a fluid power system. The pressure energy is fed into the cylinder through different control blocks known as valves. These valves are used in a hydraulic system to control the direction, pressure, flow, and other specific applications. A valve is a device that uses an external signal to release, stop or redirect the fluid flows through a system is known as a directional control valve (DCV). These external signals may be either mechanical, electrical, hydraulic, pneumatic, or fluid pilot signals. In a normally open (NO), under un-actuated conditions, the pressure port is connected to the working port or auxiliary port A. It means that fluid flows from port P to working port A. Similarly, under normally closed (NC) conditions, the pressure port is blocked and it is not connected to the working port. In this case, the fluid is not passing from the pressure port to the working port.

## 5.3  Classification of DCV

The directional control valve is classified based on fluid flow path, construction, switching position, and actuating mechanism. These classifications are mentioned here. According to the fluid flow path, the DCV is classified as,

- Check valve
- Shuttle valve
- Two-way valve
- Three-way valve
- Four-way valve.

According to construction, DCV is classified as,

- Poppet valve
- Spool valve.

According to position, DCV is classified as,

- Two-position
- Three-position valve.

According to actuation, DCV is classified as,

- Mechanical
- Electrical (Solenoid)
- Hydraulic (pilot)
- Pneumatic.

## 5.4  Check Valve

The simplest DCV is a check valve. A check valve is a mechanical device that allows fluid to flow through it in one direction and to oppose the flow in the reverse direction. It is also known as a non-return valve, one-way valve, clack valve, or reflux valve. It is a two-way valve as it has two ports. The different parts of a check valve are shown in Fig. 5.1. The ball, spring, valve seat are the components of a check valve. When fluid enters the inlet pushes the ball off the seat valve against the small force of the

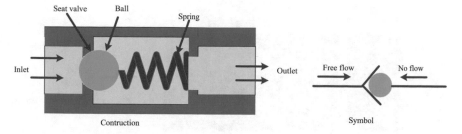

**Fig. 5.1** A ball-type check valve and its symbol

spring that continues to flow to the outlet. A very low pressure (around 15 psi) is required to open the valve in the flow position. In the opposite direction, the pressure pushes the ball in the off position so that the fluid can not pass in this direction [1].

A poppet check is also used in a fluid power system along with a spring-loaded ball check valve. For high-pressure applications, a poppet check valve is used. A poppet is used as a closing member as shown in Fig. 5.2. In a hydraulic circuit, the pilot lines are used for controlling purposes and these lines are represented by dashed lines. A certain amount of pressure is required to operate the device. Based on the pilot line, the check is classified as a pilot-to-open check valve and a pilot-to-close check valve. The pilot-to-open check valve acts as a standard check valve if there is no pressure in the pilot line as shown in Fig. 5.3.

When a specific pressure is applied to a pilot line then the pilot-to-open check permits flow in both directions.

This pilot-to-open check valve is used for locking the hydraulic cylinder from falling in case of any failure. The pilot-to-close check valve does not allow fluid flow in either direction when there is a specific pressure in the pilot line as shown in Fig. 5.4. This type of check valve is less used compared to a pilot-to-open check valve (Fig. 5.3).

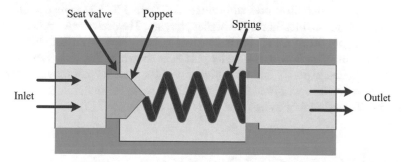

**Fig. 5.2** A poppet-type check valve

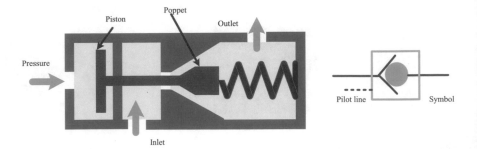

**Fig. 5.3**  A pilot-to-open check valve

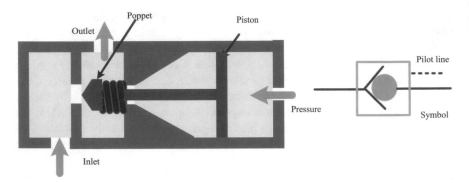

**Fig. 5.4**  A pilot-to-close check valve

## 5.5  Shuttle Valve

A shuttle valve is also known as an OR gate or double-check valve because it receives an input pressure from either port X or Y. The shuttle valve connects a pressure line (output A) to two alternative pressure sources X (P1) and Y (P2) as inputs. Outlet A receives flow from an inlet that is at a higher pressure. The pressure is applied to port X and P1 is higher than P2, then the ball moves to the right which in turn blocks port Y and port X supplies flow to the outlet port A as shown in Fig. 5.5. Similarly, the pressure is applied to port Y and P2 is higher than P1, then the ball moves to the left which in turn blocks port X and port Y supplies flow to the outlet port A as shown in Fig. 5.6. The symbol of a shuttle valve is shown in Fig. 5.7 [2].

## 5.6  Two-Way Directional Control Valves

In a two-way directional control valve, the number of ways refers to the number of connection ports. The different positions of the two-way directional control valves

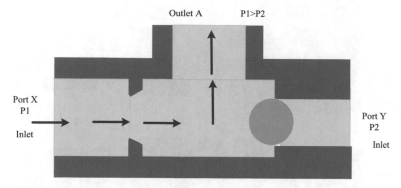

**Fig. 5.5**   A shuttle valve with port X is at higher pressure

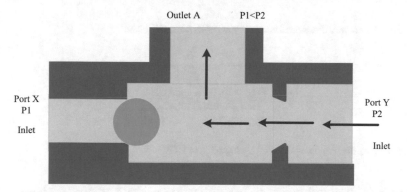

**Fig. 5.6**   A shuttle valve with port Y is at higher pressure

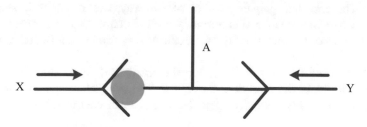

**Fig. 5.7**   A symbol of the shuttle valve

along with graphic symbols are shown in Fig. 5.8. Here, pressure port P is connected to the pump, and auxiliary port A is usually connected to the cylinder. In the top diagram of the circuit in Fig. 5.8a, the pressure is entering point P. However, the auxiliary connection port A is blocked by the spool land. This condition is known as normally closed (NC) as shown in besides graphic symbol. Again, a force is applied

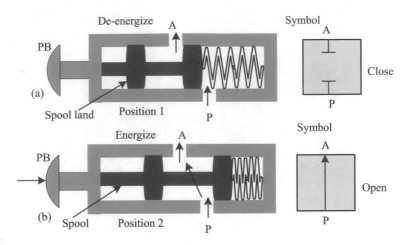

**Fig. 5.8** A two-way directional control valve with normally close

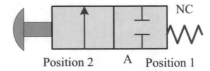

**Fig. 5.9** Two-position two-ways NC spring control DCV

to the spool land that pushes the spring and opens the fluid flow as shown in the bottom circuit in Fig. 5.8b [3, 4].

This condition is known as normally open (NO) as shown in besides graphic symbol. The complete graphic symbol of two-way, two positions is shown in Fig. 5.9. A two-way, two-position normally open directional control valve is shown in Fig. 5.10. Under normal conditions, the fluid flows from port P to A through the spool.

When pressing the push button, then the spool land closes the exhaust port. Therefore, it blocks the fluid flow through it. The complete graphic symbol of a two-position, two-way normally open (NO) directional control valve is shown in Fig. 5.11.

## 5.7  Three-Way Directional Control Valves

A three-way valve either is used to allow or block the fluid flow from the pump port to an output port. This valve is also used to allow the fluid to flow back to the tank during the pump flow is blocked. A three-way, two-position normally open spring

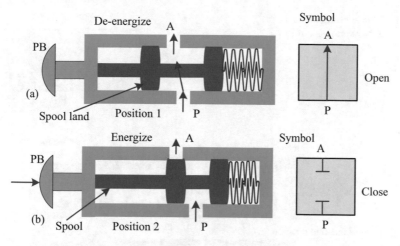

Fig. 5.10 A two-way directional control valve normally open

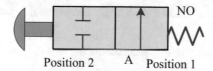

Fig. 5.11 Two-position two-way NO spring control DCV

control directional control valve is shown in Fig. 5.12. It has two fluid paths used alternatively between three ports to direct fluid to and from a single-acting cylinder.

Port A is connected to a cylinder, port P to a source of pressure, and port T is connected to a tank. Initially, the pressure port is open and fluid flows from port P to auxiliary port A and the tank T is blocked as can be seen in a symbol as shown

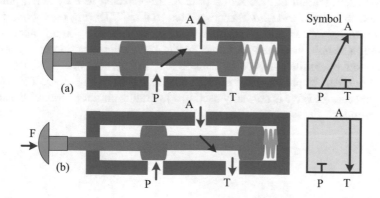

Fig. 5.12 A three-way directional control valve with normally open

**Fig. 5.13** Two-position
three-way NO spring control
DCV

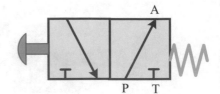

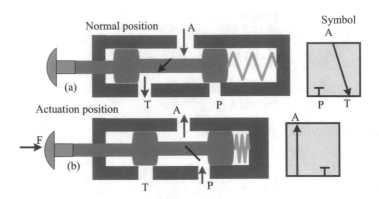

**Fig. 5.14** Two-position three-way NC spring control DCV

in Fig. 5.12a. Again, a force is applied to a push-button switch, then the spool land
pushes the spring which in turn closes the pump port and opens the tank port as can
be seen in a symbol in Fig. 5.12b. The complete graphic symbol is shown in Fig. 5.13.
This type of directional control valve is normally used in a hydraulic circuit to control
the single-acting cylinder.

A three-way, two-position normally closed (NC) spring control directional control
valve is shown in Fig. 5.14. It has two fluid paths used alternatively between three
ports to direct fluid to and from a single-acting cylinder.

In the first diagram of Fig. 5.14a, port A is connected to a cylinder, port P is
connected to a pump and port T is connected to a tank. During retraction, fluid is
coming out from the cylinder and goes to the tank. Again, a force is applied to a
push-button switch, then the spool land pushes the spring which in turn closes the
tank port and opens the pump port as can be seen in a symbol in Fig. 5.14b.

The complete graphic symbol of 3/2 DCV is shown in Fig. 5.15. This type of
directional control valve is normally used in a hydraulic circuit to control the single-
acting cylinder.

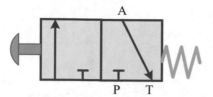

**Fig. 5.15** Two-position three-way NC spring control DCV

## 5.8 Application of 3/2 NC Directional Control Valve

A 3/2 normally closed (NC) directional control valve (DCV) is used to control the operation of a single-acting hydraulic cylinder. A lever actuation is used to operate the valve. The fixed displacement pump is run by an electric motor. The reservoir and the tank are connected with the pump and the tank. The whole circuit is drawn using Automation Studio software as shown in Fig. 5.16. Then click on the normal stimulation and activate the DCV by pressing the lever. In this case, fluid will pass through the line as indicated by the blue arrow as shown in Fig. 5.16, which in turn extends the piston of the piston. This position of the cylinder is known as extension. The state of the extension can be changed by pressing the lever. As such, when the lever of the DCV is pressed, the cylinder piston will be returned to the original

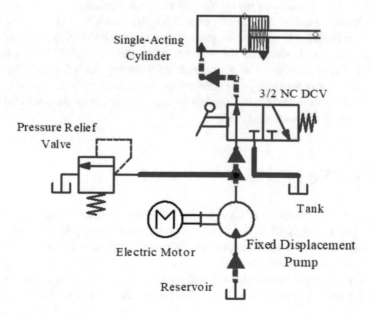

**Fig. 5.16** A single-acting cylinder with an extended position

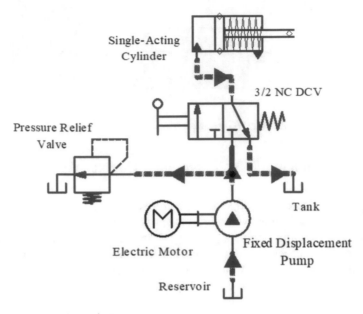

**Fig. 5.17**  A single-acting cylinder with a retracted position

position and the fluid will return to the tank as indicated by the blue arrow as shown in Fig. 5.17. This position of the cylinder is known as a retraction of the cylinder.

The pressure relief valve is usually used in a hydraulic circuit to protect the circuit by allowing an alternate path for the pump flow to go back to the tank when a preset maximum pressure is reached. This maximum pressure is slightly higher than the normal operating pressure. The pressure relief valve regulates the circuit pressure by diverting excess pressure to the tank. The circuit breaker is the analogy of the pressure relief valve. The circuit breaker operates and controls the electrical circuit under normal and abnormal conditions.

## 5.9  Four-Way Directional Control Valve

The four-way valves are used in a hydraulic circuit to control the double-acting cylinder. The four-way DCV has four connection ports namely P, T, A, and B. The ports P and T are connected to the pump and the tank, respectively. The auxiliary ports A and B are connected to the double-acting cylinder. Under the normal operating condition, port P is connected to port B and port A is connected as shown in Fig. 5.18a. This connection sequence of this DCV is used during the retraction of a double-acting cylinder.

When pressing the push-button switch i.e., under energized conditions, the pump or pressure port P is connected to the auxiliary terminal A and the auxiliary terminal

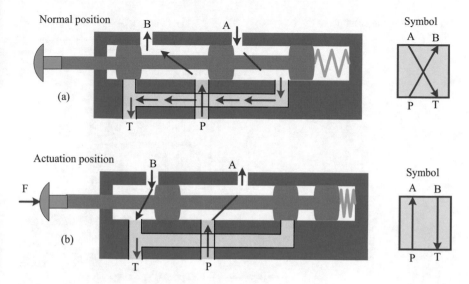

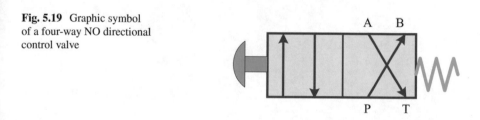

**Fig. 5.18**  A four-way directional control valve

**Fig. 5.19**  Graphic symbol of a four-way NO directional control valve

B is connected to the tank port as shown in Fig. 5.18b. This position of the DCV is used to extend the double-acting cylinder. The complete graphical symbol of a four-way DCV is shown in Fig. 5.19.

## 5.10  Application of Four-Way No Directional Control Valve

The four-way valves are used in a hydraulic circuit to control the double-acting cylinder. The four-way DCV has four connection ports namely P, T, A, and B as shown in Fig. 5.20. The fixed displacement pump is coupled to an electric motor to generate fluid flow in a system. A pressure relief valve is connected to the output of a pump to maintain the constant fluid pressure by discharging excess pressure to the tank. When pressing the lever of a 4/2 DCV, the pressure port is connected to the input of the cylinder through working port A and the fluid pressure pushes the piston of the cylinder. The cylinder output port is connected to the tank through the T port. As a result, the cylinder piston rod is coming out and this position is known

**Fig. 5.20**  A cylinder with a
4/2 DCV

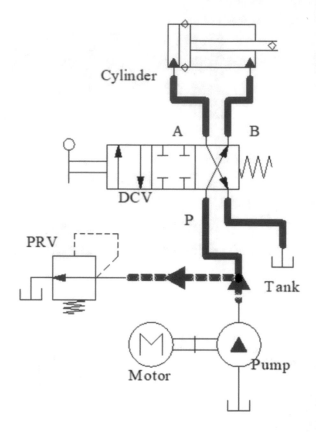

as the extended position of the cylinder as shown in Fig. 5.21. Again, pulling the
lever by clicking on it, the pressure port is connected to the working port B. The
fluid is entering the output port of the cylinder through the working port B pushes the
piston back to the original position known as the retraction position of the cylinder
as shown in Fig. 5.22.

## 5.11  Four by Three Closed Center DCV

The four-by-three closed center directional control valve is shown in Fig. 5.23. In a
central position, ports P, T, A, and B are not connected that are blocked from each
other. Therefore, the cylinder connected to working ports A and B is hydraulically
locked and can not be separated by any other external forces. As a result, the temperature of the fluid increases. This activity promotes oil oxidation, viscosity drop, and
increased system leakage. The closed center design increases the wear and shortens
the pump life. The closed center valve is suitable for use in a parallel circuit. It is
meant that the fluid flow is diving or splitting into two branches.

**Fig. 5.21** Extension position of a cylinder with 4/2 DCV

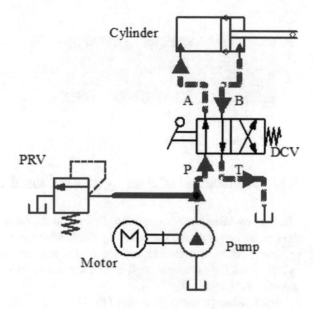

**Fig. 5.22** Retraction position of a cylinder with 4/2 DCV

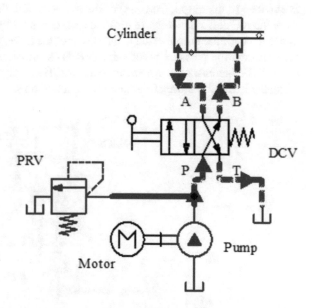

In position 1 of a graphical symbol, the cylinder is extended as the pressure port is connected to the input (cap end). Whereas in position 2, the cylinder is retracted as the pressure port is connected to the output (rod end).

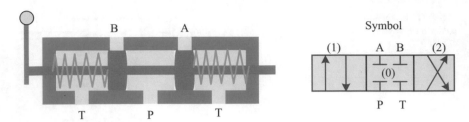

**Fig. 5.23** A closed center 4/3 DCV

## 5.12   Application of Four by Three Closed Center DCV

The four-by-three closed center directional control valve along with other necessary devices are connected as shown in Fig. 5.24. After pressing the lever of the DCV, the pressure port is connected to working port A and then to the input (cap end) of the cylinder. The fluid pressure applied forces to the cylinder which in turn extended the cylinder as shown in Fig. 5.24.

Again, when pressing the lever of the DCV, the pressure port is connected to port B and then to the output (rod end) of the cylinder. The fluid pressure applied forces from the rod end side which in turn the cylinder is retracted as shown in Fig. 5.25. When the cylinder is connected to the center of the DCV, the pump flow passes to the tank through the pressure relief valve (PRV) as shown in Fig. 5.26. This operation of the DCV increases the system temperature. This type of operation can be verified using the hydraulic trainer in a fluid power laboratory.

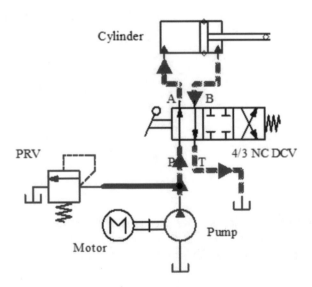

**Fig. 5.24** Extension position of a cylinder with 4/3 NC DCV

**Fig. 5.25** Retraction
position of a cylinder with
4/3 NC DCV

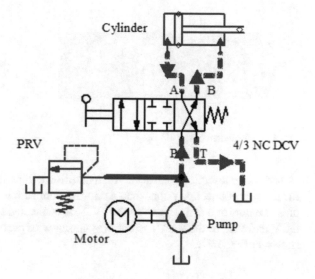

**Fig. 5.26** Cylinder
connected to the center
position of 4/3 NC DCV

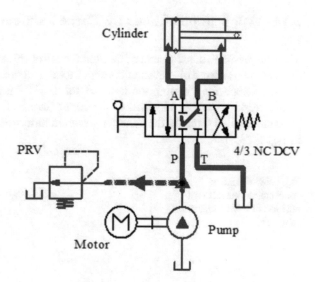

## 5.13   Four by Three Tandem Center DCV

The four-by-three tandem center directional control valve is shown in Fig. 5.27. In a tandem center position, ports the working ports A and B are blocked, but the pump port is connected to the tank port. When the cylinder is locked at the neutral position, the pump motor delivers fluid with a low pressure such as 100 psi to the valve. In this position, the pump is considered unloaded and consumed less energy when the valve is not actuated. In a tandem center, the wastage of energy is less compared to the

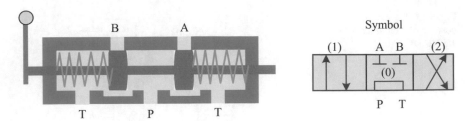

**Fig. 5.27** A tandem center 4/3 DCV

closed center direction control valve. In addition to that, the tandem center is suitable for a series connection rather than a parallel connection. In a series connection, the pressure port of a first tandem center DCV is connected to the second tandem center DCV and then to the tank. The cutaway along with ports and the graphic symbol is shown in Fig. 5.27.

## 5.14   Application of Four by Three Tandem Center DCV

The four-by-three tandem center directional control valve along with other necessary devices such as fixed displacement pump, motor, and pressure relief valve are shown in Fig. 5.28. After pressing the lever of the DCV, the pressure port is connected to the working port A and then to the input (cap end) of the cylinder. The fluid pressure applied forces to the cylinder which in turn extended the cylinder as shown in Fig. 5.28.

**Fig. 5.28** Cylinder connected to the leftmost position of a 4/3 tandem center DCV

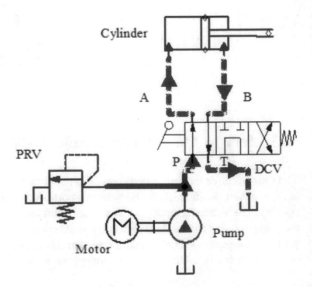

The pump pressure port P is connected to the rod end of the cylinder via the working port B. As a result, the cylinder is retracted as shown in Fig. 5.29. In addition to that, when the cylinder is locked to the tandem center, the pump supplies low-pressure fluid to the tank as shown in Fig. 5.30.

The low-pressure fluid of the first tandem center DCV is connected to the pressure port of the second DCV as shown in Fig. 5.31. When pressing the lever of the second

**Fig. 5.29** Cylinder connected to the rightmost position of a 4/3 tandem center DCV

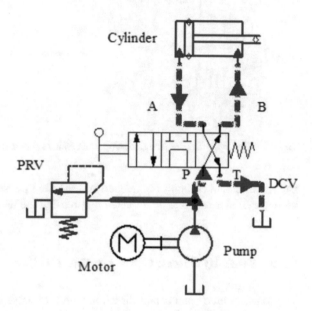

**Fig. 5.30** Cylinder connected to the center position of a 4/3 tandem center DCV

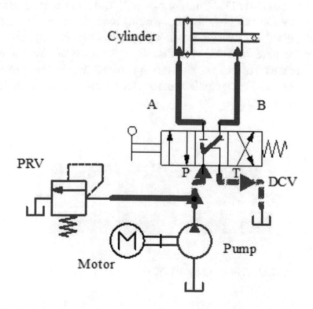

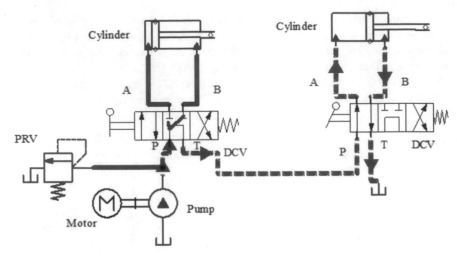

**Fig. 5.31** Two cylinders are in series using a 4/3 tandem center DCV

DCV, the second cylinder is extended and the fluid passes to the tank. Similarly, the second cylinder can also be retracted by pressing the lever.

## 5.15   Four by Three Open Center DCV

The four-by-three open center directional control valve is shown in Fig. 5.32. In an open center DCV, the four ports P, T, A, and B are connected. When the cylinder is at the center position, the working ports A and B are internally connected to the tank port T. As a result, the pump port P is blocked and all the pump flow goes back to the tank. In addition to that, the cylinder is not locked at the center position which means it extends as the working ports A and B (low-pressure ports) are connected to the tank. The graphic symbol of a 4/3 open center directional control valve is shown in Fig. 5.32.

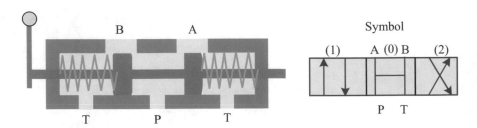

**Fig. 5.32** A open center 4/3 DCV

**Fig. 5.33** Cylinder extended
using a 4/3 open center DCV

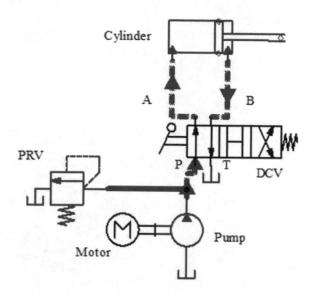

## 5.16   Application of Four by Three Open Center DCV

The four-by-three open center directional control valve along with other necessary devices such as fixed displacement pump, motor, and pressure relief valve are
connected for an application. After pressing the lever of the DCV, the pressure port
is connected to the working port A and then to the input (cap end) of the cylinder.
As a result, the cylinder is extended as shown in Fig. 5.33.

   Again, when the lever is pressing, the pressure port P is connected to the rod end
port of the cylinder through the working port B. As a result, the cylinder is retracted as
shown in Fig. 5.34. However, when the cylinder is in the center position, the cylinder
is extended due to low-pressure fluid passing through port A and entering the cap
end of the cylinder that extends slowly as shown in Fig. 5.35.

## 5.17   Four by Three Floating Center DCV

The four-by-three floating center directional control valve is shown in Fig. 5.36. In a
floating center, the working ports A and B are connected to the tank port. However,
the pressure port P is blocked and the working ports A and B are connected to tank
port T so that the cylinder can move freely that is why it is named the floating center.
Since the pump port P is blocked that is why a pressure compensated pump is required
to complete the operation. The pump produces high-pressure fluid that can pass to
the tank through the pressure relief valve and raises the heat in the system which
decreases the efficiency. Therefore, this center type of directional control valve is
used in a special application.

**Fig. 5.34** Cylinder retracted
using a 4/3 open center DCV

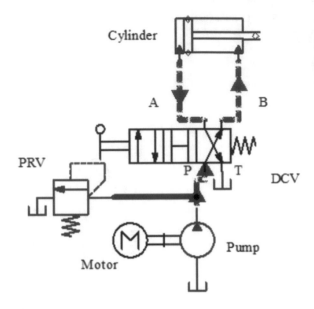

**Fig. 5.35** Cylinder extended
when is at a center position
of a 4/3 open center DCV

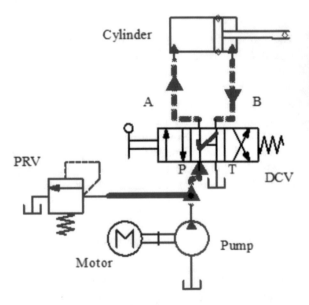

## 5.18   Application of Four by Three Floating Center DCV

The four-by-three floating center directional control valve along with other necessary devices such as fixed displacement pump, motor, and pressure relief valve are
connected for a special application. After pressing the lever of the DCV, the pressure
port is connected to the working port A and then to the input (cap end) of the cylinder.

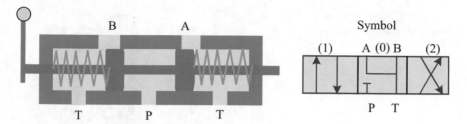

Symbol

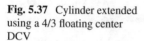

**Fig. 5.36** A floating center 4/3 DCV

**Fig. 5.37** Cylinder extended using a 4/3 floating center DCV

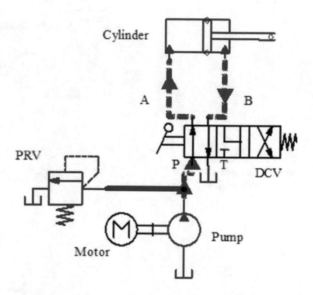

As a result, the cylinder is extended as shown in Fig. 5.37. Again, when pressing the lever of the directional control valve, the pressure port P is connected to working port B and the fluid enters into the cylinder through the rod-end port. As a result, the cylinder is retracted as shown in Fig. 5.38.

When the cylinder is in the floating center position, the working ports A and B are connected to tank port T. Since, the working ports A and B are connected to tank port T, the cylinder can move freely by applying any external forces. Hence, it is named a floating center as shown in Fig. 5.39.

## 5.19 Four by Three Regenerative Center DCV

The four-by-three regenerative center directional control valve is shown in Fig. 5.40. Regenerative means that the flow is generated from the circuit and added or supplement to the input. In a regenerative center, the pressure port P is connected to the

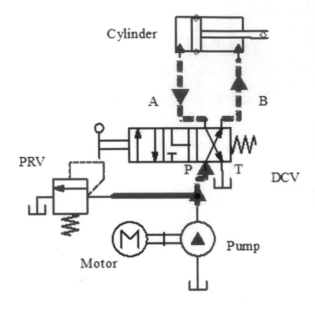

**Fig. 5.38** Cylinder retracted using a 4/3 floating center DCV

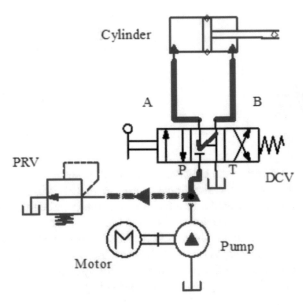

**Fig. 5.39** Cylinder is in mid-position using a 4/3 floating center DCV

working ports A and B, while the tank port is blocked. In a hydraulic circuit, a regenerative center is used when the cylinder in one direction needs two different speeds such as fast movement under no-load conditions and slow-motion under load conditions. The cutaway and the graphic symbol are shown in Fig. 5.40.

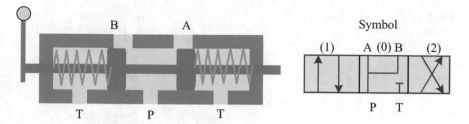

**Fig. 5.40**  A cutaway and graphic symbol of regenerative center 4/3 DCV

## 5.20  Application of Four by Three Regenerative Center DCV

The four-by-three regenerative center directional control valve along with other necessary devices such as fixed displacement pump, motor, and pressure relief valve are connected for an application in a hydraulic circuit. After pressing the lever of the DCV, the pressure port is connected to the working port A and then to the input (cap end) of the cylinder. As a result, the cylinder is extended as shown in Fig. 5.41.

Again, when pressing the lever of the directional control valve, the pressure port P is connected to working port B and the fluid enters into the cylinder through the rod-end port. As a result, the cylinder is retracted as shown in Fig. 5.42. However, when the cylinder is in the mid position, the cylinder is extended faster than the previous action as shown in Fig. 5.43. Because the return flow supplements the input of the cylinder instead of sending it to the tank.

**Fig. 5.41**  Cylinder extended using a 4/3 regenerative center DCV

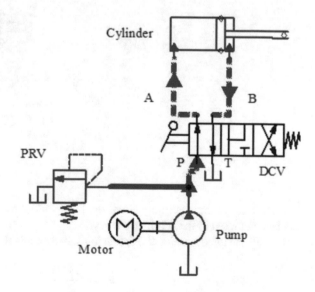

**Fig. 5.42** Cylinder retracted
using a 4/3 regenerative
center DCV

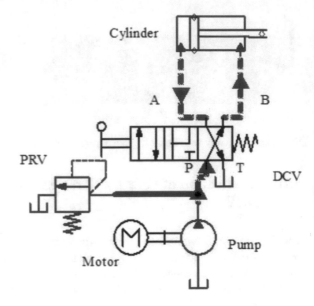

**Fig. 5.43** Cylinder retracted
using a 4/3 regenerative
center DCV

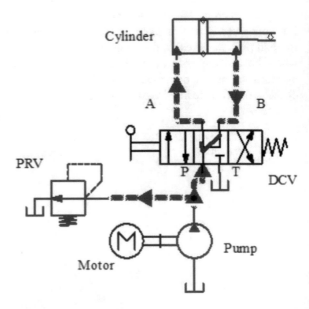

According to the continuity equation, the speed of the cylinder during extension
can be calculated in the following ways:

For US customary units, the extension speed is calculated as,

$$v_E = \frac{231Q}{A_p} \tag{5.1}$$

For the SI unit, the extension speed is calculated as,

$$v_E = \frac{Q}{1000A_p} \tag{5.2}$$

where,

$v_E$ is the speed of the cylinder during extension in m/min, in./min,
$A_p$ is the area of the cylinder piston in im$^2$, in.$^2$.

When the cylinder is connected to the mid position, the flow coming out from the rod end is added to the pump flow. As a result, the total flow is expressed as,

$$Q_t = Q + Q_R \tag{5.3}$$

where,

$Q$ is the pump flow,
$Q_R$ is the fluid flow coming out from the rod end.

Under regenerative mode, the cylinder is extended faster, and in this case, the speed is expressed as,

$$v_{E\,Regen} = \frac{231Q_t}{A_p} \tag{5.4}$$

Substituting Eq. (5.3) into Eq. (5.4) yields the extension speed in US Customary,

$$v_{E\,Regen} = \frac{231(Q + Q_R)}{A_p} \tag{5.5}$$

The extension speed in metric unit is,

$$v_{E\,Regen} = \frac{(Q + Q_R)}{1000A_p} \tag{5.6}$$

The extension speed in the regenerative mode is higher than the extension speed is in the normal mode. The rod area is lower than the piston area. Therefore, the extension speed due to regenerative mode is calculated by considering rod area as,

$$v_{E\,Regen} = \frac{231Q}{A_R} \tag{5.7}$$

$$v_{E\,Regen} = \frac{Q}{1000A_R} \tag{5.8}$$

The actual area of the rod end is the annular area and it is expressed as,

$$A_{an} = A_p - A_R \tag{5.9}$$

The expression of fluid flow coming out from the cylinder is expressed as,

$$Q_{Regen} = \frac{v_{E\,Regen}(A_p - A_R)}{231} \tag{5.10}$$

$$Q_{Regen} = 1000 \times v_{E\,Regen}(A_p - A_R) \tag{5.11}$$

The force due to regenerative mode is the net force that the cylinder is extended. It is calculated as the difference between the force applied to the cap end and the force applied to the rod end,

$$F_{Regen} = pA_p - (A_p - A_R) = pA_R \tag{5.12}$$

***Example 5.1*** A cylinder with a bore diameter of 3 in. and the rod diameter of 2 in. is attached in the mid position of the 4/3 regenerative center DCV. Calculate the extension speed with and without regenerative modes if the hydraulic circuit uses 15 GPM.

**Solution**
The value of the piston and rod areas are calculated as,

$$A_p = \pi \frac{D^2}{4} = \pi \frac{3^2}{4} = 7.07 \text{ in.}^2 \tag{5.13}$$

$$A_R = \pi \frac{d^2}{4} = \pi \frac{2^2}{4} = 3.14 \text{ in.}^2 \tag{5.14}$$

The normal extension speed is calculated as,

$$v_E = \frac{231Q}{A_p} = \frac{231 \times 15}{7.07} = 490.10 \text{ in./min} \tag{5.15}$$

The regenerative extension speed is calculated as,

$$v_{E\,Regen} = \frac{231Q}{A_R} = \frac{231 \times 15}{3.14} = 1103.50 \text{ in./min} \tag{5.16}$$

**Practice Problem 5.1**
A cylinder with a bore diameter of 60 mm and the rod diameter of 30 mm is attached in the mid position of the 4/3 regenerative center DCV. Calculate the extension speed with and without regenerative modes if the hydraulic circuit uses 30 LPM.

**Fig. 5.44** The real and
symbol of a shut-off valve

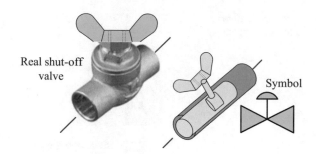

Real shut-off
valve

Symbol

## 5.21  Shut-Off Valve

The two-way valves are used either to allow fluid flow or block from the pump to a
hydraulic circuit. A manual shut-off valve is also used to allow or block fluid flow
in a system line. It is often known as a cut-off valve, lock-out valve as shown in
Fig. 5.44. This type of valve is very popular as it is very efficient and cost-effective.

The shut-off valves are used in residential, commercial, and industrial applications
for handling liquid, water, and air. A gate valve is another type of manual shut-off
valve that is used to lower and raise a gate into a hydraulic fluid stream either to
start or stop. However, this type of valve is used at home near the wall to shut off
and on the water supply. There are two types of shut-off valves are available such as
normally closed (NC) and normally open (NO). The normally open shut-off valve is
generally more common than the normally open one. Because the flow of the pipe or
process is usually not interrupted unless it is necessary to be interrupted. However,
the normally closed shut-off valve is especially useful when the subsections of a
piping system are only required for a small amount of time.

## 5.22  Actuation Methods

The method of moving the valves from one position to another position during an
operation in a fluid power circuit is known as actuation. Four basic types of actuation
are used in fluid power. These are manual, mechanical, solenoid-operated, and pilot-
operated. In manually operated DCVs, the spool is shifted by moving the lever or foot
pedal to activate the DCV. Whereas in mechanically operated, the spool is shifted by
mechanical linkages such as cam and rollers. However, in solenoid-operated DCVs,
an electric coil or a solenoid is energized by an electrical supply. As a result, it creates
a magnetic force that pulls the armature into the coil which in turn pushes the spool of
the valve. Finally, in pilot operation, the DCVs are shifted by applying a pilot signal
either hydraulic or pneumatic against a piston. When the pilot pressure is applied,
it pushes the piston to move in the required direction. The actuation methods are
shown in Fig. 5.45

**Fig. 5.45** Different
actuation methods

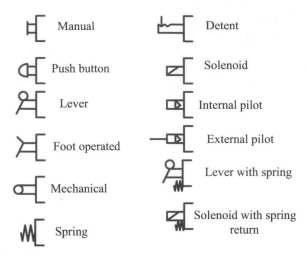

## Exercise Problems

5.1     A cylinder with a bore diameter of 2.5 in. and the rod diameter of 1.5 in. is attached in the mid position of the 4/3 regenerative center DCV. Calculate the extension speed with and without regenerative modes if the hydraulic circuit uses 12 GPM.

5.2     A cylinder with a bore diameter of 55 mm and the rod diameter of 35 mm is attached in the mid position of the 4/3 regenerative center DCV. Find the extension speed with and without regenerative modes if the hydraulic circuit uses 15 LPM.

5.3     A cylinder with a bore diameter of 4.5 in. and the rod diameter of 3.5 in. is attached in the mid position of the 4/3 regenerative center DCV. Calculate the maximum extension forces with and without regenerative modes, if the hydraulic circuit pressure is 1200 psi.

# References

1. A. Esposito, *Fluid Power with Applications*, 7th edn. (Pearson New International Education, The United States of America, 2014), pp. 1–648
2. F. Don Norvelle, *Fluid Power Technology*, 1st edn. (Delmar, a Division of Thomson Learning, The United States of America, 1995), pp. 1–649
3. J.A. Sullivan, *Fluid Power-Theory and Applications*, 3rd edn. (Prentice-Hall, New Jersey, The United States of America, 1989), pp. 1–528
4. J.L. Johnson, *Introduction to Fluid Power*, 1st edn. (Delmar, a Division of Thomson Learning, The United States of America, 2002), pp. 1–502

# Chapter 6
# Hydraulic Motors

## 6.1 Introduction

The actuators are classified as motors and cylinders that are used in the hydraulic systems to utilize the pump output as their input power. The energy delivered to the hydraulic system by the pump is generally utilized to run the linear device (cylinder) or rotary device (motor). The pump converts mechanical power from an electric motor into fluid power. However, the motors perform opposite functions of the pump. The hydraulic motor converts fluid power into mechanical power in the form of rotation (angular displacement) of the shaft by converting the fluid pressure into torque. This rotation is continuous in the case of hydraulic motor and it exerts limited rotation in the case of torque motor. In general, the hydraulic motor produces rotation of shaft and torque when high-pressure fluid is supplied. This chapter will discuss different types of hydraulic motors, working principles, torque, speed, efficiency, graphical symbols, and applications.

## 6.2 Classification of Hydraulic Motor

The hydraulic motor is designed for working pressure at both sides of the motor. The motor can be classified as fixed and variable displacement. Increasing the displacement of a motor decreases the speed and increases the torque. Similarly, decreasing the displacement increases the speed and decreases torque. Most of the hydraulic motors are positive displacement types. Hydraulic motors are classified into gear, vane, and piston types. However, gear, piston, and vane motors are considered into the fixed displacement group.

The gear motor consists of two gears one is known as drive gear attached with the stator and the other is known as driven gear that is attached to the output shaft as shown in Fig. 6.1.

**Fig. 6.1** Schematics of gear
and vane motors

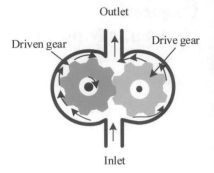

Outlet

Driven gear                Drive gear

Inlet

Both gears have the same number of teeth. The high-pressure fluid enters into the inlet of the motor. Then it flows around the surface between the gear teeth and the wall housing and to the outlet of the motor. The surface area between the gear teeth and wall housing has less resistance, so that fluid passes easily. Gear motor has low weight, a simple design, and medium pressure. It is used in agricultural machinery to drive conveyor belts, distribution plates, etc.

The vane motor is a positive displacement motor that develops output torque on its shaft by allowing fluid to act on the vanes. This motor consists of vanes, housing with an eccentric bore that runs a rotor with associated vanes that comes in and out. This action creates an unbalanced force of the pressurized fluid on the vanes. As a result, the rotor of the motor rotates in one direction. The main concern is about the correct design of a vain tip to the motor housing contact point. These types of contact points need to be sealed properly to reduce wear and metal-to-metal contact that increases the efficiency. The vane motor has high operating efficiency but not as much as the piston motor. However, the vane motor is less costly than the piston motor. The schematic of the vane motor is shown in Fig. 6.2.

The piston motor is available in low-speed high torque (LSHT) and high-speed low torque (HSLT). The motor bore contains the number of pistons that are reciprocating to each other. The pressurized fluid enters through the inlet that pressed the series of pistons inside the barrel with a fixed angle with the swashplates. These pistons push against this angle which creates the rotation of the swashplate that is connected

**Fig. 6.2** Schematics of vane
motor

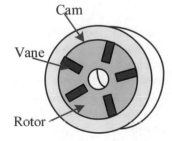

Cam

Vane

Rotor

**Fig. 6.3** Schematics of a
radial piston motor

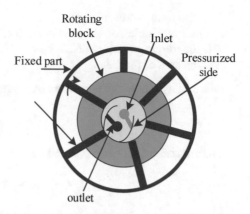

to the output shaft of the motor. The schematic of a radial piston motor is shown in
Fig. 6.3.

## 6.3 Motor Torque

The torque is one of the key features of a hydraulic motor. The output torque is
produced by the pressure difference between incoming and outgoing hydraulic fluid
and displacement during one revolution of the shaft. A hydraulic vane motor with a
cam, rotor, and vane is shown in Fig. 6.4. The outer part is known as a cam and the
inner part is known as a rotor. The rotor rotates when the fluid pressure acts on the
vane. For a single rotation of a vane, consider the following [1, 3]:

$L$ is the length of the vane in m, in.,
$R_c$ is the cam radius in m, in.,

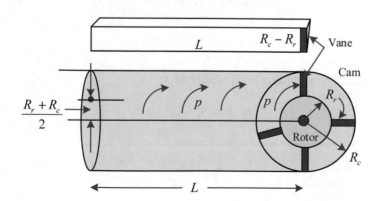

**Fig. 6.4** A simple hydraulic motor

$R_r$ is the rotor radius in m, in.,
$p$ is pressure acting on the vane in Pa, psi,
$T$ is the torque capacity in N.m, in.lb,
$A$ is the area of the vane.

Due to a hydraulic pressure acted on a vane, the force is calculated as,

$$F = pA \tag{6.1}$$

The area of the vane is calculated as,

$$A = (R_c - R_r) \times L \tag{6.2}$$

Substituting Eq. (6.2) into Eq. (6.1) yields,

$$F = p(R_c - R_r)L \tag{6.3}$$

The torque capacity of a hydraulic motor is calculated by using the force and the distance of the center of rotation. From Fig. 6.1, the distance of the center of rotation ($d_r$) is calculated as,

$$d_r = \frac{R_c + R_r}{2} \tag{6.4}$$

The torque capacity of a hydraulic motor is calculated as,

$$T = F \times d_r \tag{6.5}$$

Substituting Eqs. (6.3) and (6.4) into Eq. (6.5) yields,

$$T = p(R_c - R_r)L \times \frac{R_c + R_r}{2} \tag{6.6}$$

$$T = \frac{pL}{2}(R_c^2 - R_r^2) \tag{6.7}$$

The volumetric displacement is calculated as,

$$v_d = \pi (R_c^2 - R_r^2) \times L \tag{6.8}$$

Substituting Eq. (6.8) into Eq. (6.7) yields,

$$T = \frac{p\, v_d}{2\pi} \tag{6.9}$$

From Eq. (6.9), it is seen that the torque capability of a hydraulic vane motor can be calculated if other parameters are given.

***Example 6.1*** The displacement of a hydraulic motor is given by 1.5 in.$^3$/rev. Calculate the output torque if the maximum pressure of the system is found to be 1000 psi.

**Solution**
The value of the output torque is calculated as,

$$T = \frac{p v_d}{2\pi} = \frac{1000 \times 1.5}{2\pi} = 238.73 \, \text{in.lb} \tag{6.10}$$

**Practice Problem 6.1**
The displacement of a hydraulic motor is given by 35 cm$^3$/rev. Calculate the output torque if the m's maximum pressure is found to be 15,000 kPa.

## 6.4 Motor Speed

The speed of a hydraulic motor depends on the amount of flow coming out from the pump and the displacement of the motor. The speed of the motor is directly proportional to the flow rate and inversely proportional to the motor displacement. Mathematically, it can be expressed as [2],

$$N = \frac{Q}{V_m} \tag{6.11}$$

$$Q = V_m N \tag{6.12}$$

where,

$N$    is the motor speed in rev/min,
$V_m$   is the motor displacement in in.$^3$/rev,
$Q$    is the flow rate in in.$^3$/min, m$^3$/min.

For the US customary unit, use 1 GPM $= 231$ in.$^3$/min, then Eq. (6.12) is modified as,

$$Q = \frac{V_m N}{231} \tag{6.13}$$

However, in a metric unit, Eq. (6.12) is revised as,

$$Q = \frac{V_m N}{1000} \tag{6.14}$$

**Example 6.2** A hydraulic motor rotates at 800 rpm with a 3 in.$^3$/rev. Calculate the flow rate by assuming 100% efficiency.

**Solution**
The value of the flow rate is calculated as,

$$Q = \frac{V_m N}{231} = \frac{3 \times 800}{231} = 10.39 \, \text{GPM} \tag{6.15}$$

**Example 6.3** A hydraulic motor rotates at 500 rpm with a 35 cm$^3$/rev. Calculate the flow rate by assuming 100% efficiency.

**Solution**
The value of the flow rate is calculated as,

$$Q = \frac{V_m N}{1000} = \frac{35 \times 500}{1000} = 17.50 \, \text{LPM} \tag{6.16}$$

**Practice Problem 6.2**
A hydraulic motor rotates at 600 rpm with a 2.5 in.$^3$/rev. Calculate the flow rate by assuming 100% efficiency.

**Practice Problem 6.3**
A hydraulic motor rotates at 700 rpm with a 45 cm$^3$/rev. Calculate the flow rate by assuming 100% efficiency.

## 6.5  Motor Power

The motor converts fluid flow from the pump output into a rotational speed at its shaft. Later on, this rotational speed is used to do work. In this case, the motor shaft needs to couple with a load shaft to transmit the power. To derive the power of the motor, the following unit conversions need to be included [3].

The unit in rad per second (rad/s) is converted in the following ways,

$$\text{rad/s} \times \text{s/min} \times \text{rev/rad} = \text{rpm} \tag{6.17}$$

But, the radian and revolution is related as,

$$2\pi \, \text{rad} = 1 \, \text{rev} \tag{6.18}$$

Substituting Eq. (6.18) and 1 min = 60 s in Eq. (6.17) yields,

$$N \text{ rad/s} \times \text{s/60s} \times \frac{1}{2\pi} \text{rev/rad} = N \text{ rpm} \tag{6.19}$$

$$N \text{ rad/s} = \frac{2\pi}{60} N \text{ rpm} \tag{6.20}$$

The shaft power of the motor is equal to the product of the torque ($T$) and the rotational speed ($N$). Mathematically, it is expressed as,

$$P_s = T(\text{ft-lb}) \times N(\text{rad/s}) = \frac{2\pi}{60} T_A \times N(\text{rpm}) = \frac{T \times N}{60/2\pi} = \frac{T \times N}{9.549} \text{ W} \tag{6.21}$$

Using the conversion factor 1 HP = 550 ft-lb/s = 746 N-m/s (W) to modify Eq. (6.21),

$$P_s = \frac{T \times N}{9.549 \times 550} \tag{6.22}$$

$$P_s = \frac{T \times N}{5252} \text{ HP} \tag{6.23}$$

where the torque ($T$) is in ft-lb. Then it is converted to in-lb and Eq. (6.23) is modified as,

$$P_s = \frac{T \times N}{5252 \times 12} \text{ HP} \tag{6.24}$$

$$P_s = \frac{T \times N}{63025.35} \text{ HP} \tag{6.25}$$

In general, the output power of the motor is,

$$P_s = \frac{T \times N}{63025} \text{ HP} \tag{6.26}$$

In the metric unit, from Eq. (6.21), the output in kilowatts is expressed as,

$$P_s = \frac{T \times N}{9.549} \text{ W} = \frac{T \times N}{9.549 \times 1000} \text{ kW} \tag{6.27}$$

$$P_s = \frac{T \times N}{9550} \text{ kW} \tag{6.28}$$

**Example 6.4** A hydraulic motor rotates at 500 rpm to drive a load of 150 ft.lb torque. Calculate output power in horsepower.

**Solution**

The value of the torque is converted to in.lb as,

$$T = 150 \times 12 = 1800\,\text{in.lb}$$

The output power is calculated as,

$$P_s = \frac{TN}{63025} = \frac{1800 \times 500}{1000} = 900\,\text{HP} \tag{6.29}$$

***Example 6.5*** A hydraulic motor rotates at 400 rpm to drive a load of 800 N.m torque. Determine the output power in kilowatt.

**Solution**

The output power is calculated as,

$$P_s = \frac{TN}{9550} = \frac{800 \times 400}{9550} = 33.51\,\text{kW} \tag{6.30}$$

**Practice Problem 6.4**

A hydraulic motor rotates at 600 rpm to drive a load of 185 ft.lb torque. Calculate output power in horsepower.

**Practice Problem 6.5**

A hydraulic motor rotates at 500 rpm to drive a load of 600 N-m torque. Determine the output power in kilowatt.

## 6.6   Motor Efficiency

The motor efficiency is calculated to know how the fluid input horsepower is converted to the useful output brake horsepower. The efficiency of the motor is classified as volumetric, mechanical and overall efficiency. In a motor, mechanical and volumetric losses are considered. Mechanical losses occurred due to wear and friction. Whereas the volumetric losses occurred due to leakage. The volumetric efficiency is defined as the ratio of theoretical flow rate ($Q_T$) motor should consume to the actual flow rate ($Q_A$) consumed by the motor. Mathematically, the volumetric efficiency is expressed as [4],

$$\eta_v = \frac{Q_T}{Q_A} \tag{6.31}$$

The theoretical flow rate of a motor is related to the motor displacement and the rpm. From Eq. (6.13), this relation is expressed as,

$$Q_T = \frac{V_m N}{231} \tag{6.32}$$

Substituting Eq. (6.32) into Eq. (6.32) yields,

$$\eta_v = \frac{\frac{V_m N}{231}}{Q_A} \tag{6.33}$$

$$\eta_v = \frac{V_m N}{231 \, Q_A} \tag{6.34}$$

As there are wear and friction losses in a motor, the actual torque ($T_A$) is always less than the theoretical torque ($T_T$). The mechanical efficiency is defined as the ratio of the actual torque to the theoretical torque. Mathematically, the volumetric efficiency is expressed as,

$$\eta_m = \frac{T_A}{T_T} \tag{6.35}$$

From Eq. (6.9), the theoretical torque is calculated as,

$$T_T = \frac{p \, v_d}{2\pi} \tag{6.36}$$

Due to frictional losses, the actual torque is always less than the theoretical torque. From Eq. (6.26), the expression of actual torque is written as,

$$T_A = \frac{P_s}{N} \times 63025 \tag{6.37}$$

where,

$T$   is in lb-ft,
$N$   is in rpm,
$P_s$   is in HP.

Similarly, from Eq. (6.28), the actual torque in the SI unit is expressed as,

$$T_A = \frac{P_s}{N} \times 9550 \tag{6.38}$$

where,

$T$    is in N-m,
$N$    is in rpm,
$P_s$    is in kW.

Substituting equations ((6.36) and (6.37) into Eq. (6.35) yields,

$$\eta_m = \frac{\frac{P_s}{N} \times 63025}{\frac{p \, v_d}{2\pi}} \tag{6.39}$$

$$\eta_m = 395997.75 \frac{P_s}{N \, p \, v_d} \tag{6.40}$$

The overall efficiency of a motor is defined as the output horsepower power (OHP) available at the shaft to the hydraulic horsepower (HHP) applied as an input. Mathematically, it is expressed as,

$$\eta_o = \frac{OHP}{HHP} = \eta_v \times \eta_m \tag{6.41}$$

The power is defined as the work done ($W$) per unit time ($t$) and it is expressed as,

$$P = \frac{W}{t} = \frac{F \times d}{t} \tag{6.42}$$

Replacing $d/t$ with the velocity, $v$ in Eq. (6.42) yield,

$$P = F \times v \tag{6.43}$$

Again, substituting $F = pA$ in Eq. (6.43) yields,

$$P = pA \times v \tag{6.44}$$

According to the continuity equation, substituting $Q = Av$ in Eq. (6.44) yields,

$$P = pQ \tag{6.45}$$

where $p$ is the fluid pressure in psi or lb/in.$^2$ and $Q$ is the flow rate in in.$^3$/min. The unit of the power will be in.lb/min.

Let us consider 1 hp is equal to 396,000 in.lb/min and 1gpm is equal to 231 in.$^3$/min. Therefore, multiplying by 231 to flow rate in gpm and dividing by 396,000 to Eq. (6.45) to convert in horsepower (HP). This simplification can be written as,

$$HHP = \frac{pQ \times 231}{396000} \tag{6.46}$$

Equation (6.46) reduces as,

$$HHP = \frac{pQ}{1714} \qquad (6.47)$$

In the SI unit, the unit of pressure is Pa ($N/m^2$) and the flow rate is in $m^3/s$. Therefore, the unit of power will be N.m/s that is known as watt (W).

One litre is equal to 1000 cubic centimetres. Therefore, the conversion factor lpm to $m^3/min$ is expressed as,

$$1L = 1000cc = 1000 \times \frac{1}{100} \times \frac{1}{100} \times \frac{1}{100} \, m^3 \qquad (6.48)$$

$$1L = \frac{1}{1000} \, m^3 \qquad (6.49)$$

$$1 \, m^3 = 1000 \, L \qquad (6.50)$$

In Eq. (6.50), dividing both sides by second (s) yields,

$$1 \, m^3/s = 1000 \, L/s \qquad (6.51)$$

Converting second to minutes in the right side of Eq. (6.51) yields,

$$1 \, m^3/s = 1000 \, L/s/60 = 1000 \times 60 \, L/min \qquad (6.52)$$

In the SI unit, Eq. (6.45) reduces to,

$$P_H = \frac{pQ}{60000} \qquad (6.53)$$

where the pressure is in kPa, the flow rate is in lpm and the hydraulic power is in kW.

The overall efficiency is expressed as the output power at the shaft to the hydraulic power input to the motor and it is expressed as,

$$\eta_o = \frac{P_s}{P_H(HHP)} \qquad (6.54)$$

Substituting Eqs. (6.23) and (6.47) into Eq. (6.54) yields,

$$\eta_o = \frac{\frac{T \times N}{5252}}{\frac{p \times Q}{1714}} \qquad (6.55)$$

Equation (6.55) reduces to,

$$\eta_o = \frac{1714}{5252} \frac{T \times N}{p \times Q} \tag{6.56}$$

$$\eta_o = 0.326 \frac{T \times N}{p \times Q} \tag{6.57}$$

where,

$T$   is in lb-ft,
$N$   is in rpm,
$p$   is in psi,
$Q$   is in gpm.

**Example 6.6** A 500 rpm hydraulic motor operates at 1500 psi in a system. The system uses a 14 gpm flow rate and exerts a torque of 25 lb-ft. Calculate the overall efficiency.

**Solution**
The overall efficiency is calculated as,

$$\eta_o = 0.326 \frac{T \times N}{p \times Q} = 0.326 \frac{25 \times 500}{1500 \times 14} = 19.4\% \tag{6.58}$$

**Example 6.7** The volumetric displacement of a 600 rpm hydraulic motor is found to be 1.5 in.³/rev. Calculate the theoretical flow rate. Also, calculate the volumetric efficiency if the actual flow rate is 5.65 gpm.

**Solution**
The theoretical flow rate is calculated as,

$$Q_T = V_d \times N = 1.5 \times 600 = 900 \, \text{in.}^3/\text{min} \tag{6.59}$$

$$Q_T = \frac{900}{231} = 3.90 \, \text{gpm} \tag{6.60}$$

The volumetric efficiency is calculated as,

$$\eta_v = \frac{Q_T}{Q_A} = \frac{3.90}{5.65} = 68.95\% \tag{6.61}$$

**Example 6.8** The volumetric displacement of a 1600 rpm hydraulic motor is found to be 30 cm³/rev. The maximum pressure of the system is found to be 15,000 kPa. Determine the theoretical torque. Also, calculate the actual torque and mechanical efficiency if the output power of the motor is 10 kW.

**Solution**

The volumetric displacement in $m^3/rev$ is calculated as,

$$V_d = 30\,\text{cm}^3/\text{rev} = \frac{30}{100 \times 100 \times 100} = 3 \times 10^{-5}\,\text{m}^3/\text{rev} \quad (6.62)$$

The theoretical torque is determined as,

$$T_T = \frac{p\,v_d}{2\pi} = \frac{15000000 \times 3 \times 10^{-5}}{2\pi} = 71.62\,\text{N-m} \quad (6.63)$$

The value of the actual torque is calculated as,

$$T_A = \frac{P_s}{N} \times 9550 = \frac{10 \times 9550}{1600} = 59.69\,\text{N-m} \quad (6.64)$$

The mechanical efficiency is calculated as,

$$\eta_m = \frac{T_A}{T_T} = \frac{59.69}{71.62} = 83.34\% \quad (6.65)$$

**Practice Problem 6.6**

An 800 rpm hydraulic motor operates at 1200 kPa in a system. The system uses a 15 lpm flow rate and exerts a torque of 25 N-m. Calculate the overall efficiency.

**Practice Problem 6.7**

The volumetric displacement of a 600 rpm hydraulic motor is found to be 2.5 in.$^3$/rev. Calculate the theoretical flow rate. Also, calculate the volumetric efficiency if the actual flow rate is 5.65 gpm.

**Practice Problem 6.8**

The volumetric displacement of a 1200 rpm hydraulic motor is found to be 40 cm$^3$/rev. The maximum pressure of the system is found to be 25,000 kPa. Determine the theoretical torque. Also, calculate the actual torque and mechanical efficiency if the output power of the motor is 12 kW.

## 6.7 Motor Graphic Symbols

Figure 6.5 represents the components of the hydraulic motor graphic symbol. The circle represents the energy converter of the motor and the solid triangle inside the circle represents the direction of hydraulic fluid flow and direction of rotation.

If the fluid enters through the working port A, then the rotation of the motor will be in the clockwise direction. Whereas the fluid enters through the working port

**Fig. 6.5**  Components of the graphic symbol of hydraulic motor

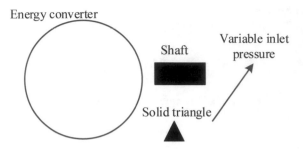

B, the motor will be in the anticlockwise direction. These rotations are shown in Fig. 6.6. The graphic symbols of fixed-displacement unidirectional and bidirectional hydraulic motors are shown in Fig. 6.7 [1].

The variable displacement bi-directional and limited rotation bi-directional hydraulic motors are shown in Fig. 6.8.

**Exercise Problems**

6.1  The displacement of a hydraulic motor is given by 45 cm³/rev. Calculate the output torque if the m's maximum pressure is found to be 10,000 kPa.

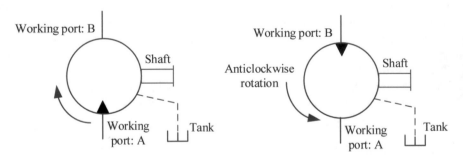

**Fig. 6.6**  Clockwise and anticlockwise directions of hydraulic motor

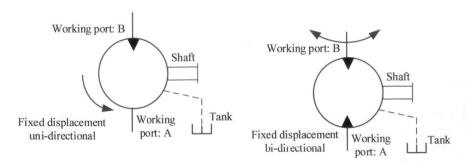

**Fig. 6.7**  Uni-directional and bi-directional motors

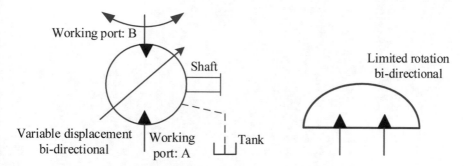

**Fig. 6.8** Variable displacement bi-directional and limited rotation bi-directional motors

6.2    A hydraulic motor rotates at 400 rpm with a 3.5 in.$^3$/rev. Calculate the flow rate by assuming 100% efficiency.

6.3    A hydraulic motor rotates at 800 rpm with a 55 cm$^3$/rev. Calculate the flow rate by assuming 100% efficiency.

6.4    A hydraulic motor rotates at 800 rpm to drive a load of 125 ft.lb torque. Calculate output power in horsepower.

6.5    A hydraulic motor rotates at 600 rpm to drive a load of 400 N-m torque. Determine the output power in kilowatt.

6.6    A 500 rpm hydraulic motor operates at 1100 kPa in a system. The system uses a 12 lpm flow rate and exerts a torque of 20 N-m. Calculate the overall efficiency.

6.7    The volumetric displacement of a 400 rpm hydraulic motor is found to be 3.5 in.$^3$/rev. Calculate the theoretical flow rate. Also, calculate the volumetric efficiency if the actual flow rate is 5.55 gpm.

6.8    The volumetric displacement of a 1400 rpm hydraulic motor is found to be 45 cm$^3$/rev. The maximum pressure of the system is found to be 23,000 kPa. Determine the theoretical torque. Also, calculate the actual torque and mechanical efficiency if the output power of the motor is 14 kW.

# References

1. J.L. Johnson, *Introduction to Fluid Power*, First Edn. (Delmar, a division of Thomson Learning, The United States of America, 2002), pp. 1–502
2. F. Don Norvelle, *Fluid Power Technology*, First Edn. (Delmar, a division of Thomson Learning, The United States of America, 1996), pp. 1–649
3. A. Esposito, *Fluid Power with Applications*, Seventh Edn. (Pearson New International Education, The United States of America, 2014), pp. 1–648,
4. J.A. Sullivan, *Fluid Power-Theory and Applications*, Third Edn. (Prentice-Hall New Jersey, The United States of America, 1989), pp. 1–528

# Chapter 7
# Hydraulic Cylinders

## 7.1 Introduction

Hydraulic cylinder converts fluid flow under pressure into linear mechanical force. The magnitude of the mechanical force depends upon the amount of flow rate, the pressure drop across the cylinder, and its overall efficiency. Hydraulic cylinders extend and retract the piston rod that provides the push or pull force to move the external load in a straight path. The cylinder is classified as ram, single-acting, double-acting, telescopic. The ram cylinder is a special type of single-acting cylinder that has the same rod diameter as the piston. This type of cylinder is mostly mounted vertically and it retracts by the force of the load due to gravity. Ram cylinders are used for elevators and jacking purposes. The single-acting cylinder uses different diameters of the piston and rod. It has one port where fluid enters and extends the piston. After extension, the cylinder retracts due to spring pressure. Double-acting is classified as double-acting single rod, double-acting double rods, and tandem.

This chapter will discuss different types of hydraulic motors, working principles, torque, speed, efficiency, graphical symbols, and applications.

## 7.2 Construction of Hydraulic Cylinder

The construction of the hydraulic cylinder is shown in Fig. 7.1. The main parts of the cylinder are barrel, piston, rod, rod end port, blind end, or cap end port. The barrel is the tube portion of the cylinder that provides desired bore diameter during extension and retraction of the cylinder with zero leakage. It also maintains adequate strength for the safe operation of the specified system operating pressure. The two ends of the barrel are closed by the parts known as the head or rod and cap or blind. The head closes the end of the cylinder through which the piston and rod assembly can place or enter. Whereas the cap closes the opposite end by proper welding, the end without the rod.

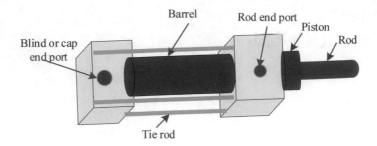

**Fig. 7.1**  Schematics of a hydraulic cylinder

## 7.3   Single-Acting Cylinder

A single-acting cylinder has one working at the blind end and the other side has a vent port as shown in Fig. 7.2. The forward motion (extension) of the cylinder is obtained by supplying pump flow to the working port. The backward motion (retraction) is obtained due to the spring pressure. The single-acting cylinders are used where the force is required only in one direction such as clamping, feeding, sorting, locking, ejecting, braking, etc. [1].

The single-acting cylinders with and without spring are shown in Fig. 7.3.

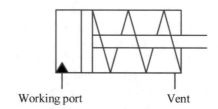

**Fig. 7.2**  Symbol of a hydraulic cylinder

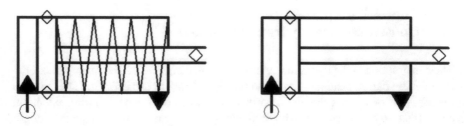

**Fig. 7.3**  Symbols of hydraulic cylinders in automation studio

**Fig. 7.4** Symbols of double-acting hydraulic cylinders in automation studio

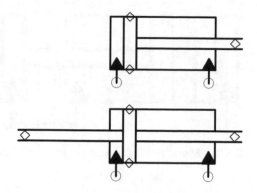

## 7.4 Double-Acting Cylinder

Two working ports are equipped with a double-acting cylinder. One working port on the piston side and the other on the rod side as shown in Fig. 7.4. The double-acting cylinders are classified as a double-acting cylinder with a single-rod and a double-acting cylinder with double rods placed on both sides.

Pump flow is supplied to the cap end and the cylinder extends. During this extension, fluid returns to the tank as the rod end port is connected to the tank. The rod extension is slower due to the larger area but it provides a larger force for the load. To retract the cylinder, pump flow is supplied to the rod end working port and the fluid returns to the tank as the cap end is connected to the tank. During retraction, the rod returns faster as the area for the fluid is smaller but is provides a smaller force. A double-acting double rods cylinder is used in special applications where the work is done by both ends of the cylinder. In addition to that, this type of cylinder provides equal force and equal speed in both directions [2].

## 7.5 Double-Acting Tandem Cylinder

A double-acting tandem cylinder has two pistons with a common rod. It is also known as a combination cylinder. In the Automation Studio software, a double-acting tandem cylinder is formed by connecting two double-acting cylinders in series as shown in Fig. 7.5.

The greater force is obtained by using a double-acting tandem cylinder where the space is insufficient to cylinder bore size. However, a high volume of fluid is required to drive the cylinder.

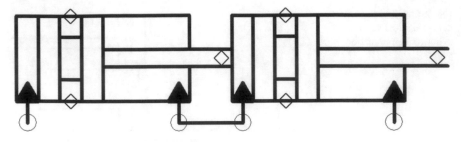

**Fig. 7.5** Symbols of double-acting tandem or combination cylinder

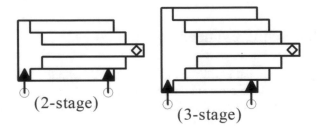

**Fig. 7.6** Symbols of double-acting tandem or combination cylinder

## 7.6　Telescopic Cylinder

The telescopic cylinder is usually single-acting and double-acting consists of series segments called sleeves. There are four to five sleeves in the cylinder and each sleeve is placed on the previous sleeve. These sleeves are expanded one by one when a larger stroke is required. The maximum force is obtained when the sleeves are at the retraction position or the collapsed position. The two-stage and three-stage double-acting telescopic cylinders are shown in Fig. 7.6.

Telescopic cylinders are used in the construction truck to unload the materials. It is also used in an amusement park to provide different stroke lengths in the rider.

## 7.7　Cylinder Force

The cylinder generates linear mechanical force when the pressure is applied to the cap end side as shown in Fig. 7.7. This force depends on the area of the piston $(A_p)$ and the system fluid pressure. During extension, the fluid pressure acts on the entire area of the piston. According to Pascal law, the force during the extension of the cylinder is written as,

$$F_e = p \times A_p \tag{7.1}$$

**Fig. 7.7** Symbols of a cylinder during extension

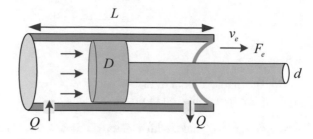

The area of the piston with a diameter $D$ is expressed as,

$$A_p = \frac{\pi D^2}{4} \qquad (7.2)$$

where,

$p$ is the fluid pressure in psi or Pa,
$D$ is the diameter in m or in,
$A_p$ is the area of the piston in $m^2$ or $in^2$.

During the retraction, the fluid enters through the rod end and the fluid pressure acts on the smaller annulus area ($A_{an}$) as shown in Fig. 7.8. This annular area is expressed as,

$$A_{an} = A_p - A_r \qquad (7.3)$$

The force due to retraction of the cylinder is written as,

$$F_r = p \times A_{an} \qquad (7.4)$$

Substituting Eq. (7.3) into Eq. (7.4) yields,

$$F_r = p \times (A_p - A_r) \qquad (7.5)$$

The area of the rod with a diameter $d$ is expressed as,

**Fig. 7.8** Symbols of a cylinder during retraction

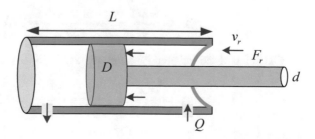

$$A_r = \frac{\pi d^2}{4} \tag{7.6}$$

where,

    $d$ is the diameter in m or in,
    $A_r$ is the area of the rod in m$^2$ or in$^2$.

***Example 7.1***  A cylinder with a piston diameter of 2.5 in and a rod diameter of 1.5 in is used in a hydraulic system that has a pressure of 1500 psi. Calculate the extension force and retraction force of the cylinder.

**Solution**

The value of the piston area is calculated as,

$$A_p = \frac{\pi D^2}{4} = \frac{\pi \times 2.5^2}{4} = 4.91 \, \text{in}^2 \tag{7.7}$$

The value of the rod area is calculated as,

$$A_r = \frac{\pi d^2}{4} = \frac{\pi \times 1.5^2}{4} = 1.77 \, \text{in}^2 \tag{7.8}$$

The value of the extension force is calculated as,

$$F_e = p \times A_p = 1500 \times 4.91 = 7365 \, \text{lb} \tag{7.9}$$

The value of the retraction force is calculated as,

$$F_r = p \times (A_p - A_r) = 1500 \times (4.91 - 1.77) = 4710 \, \text{lb} \tag{7.10}$$

***Example 7.2***  A cylinder with a piston diameter of 60 mm in and a rod diameter of 50 mm is used in a hydraulic system that has a pressure of 2000 Pa. Determine the extension force and retraction force of the cylinder.

**Solution**

The value of the piston area is calculated as,

$$A_p = \frac{\pi D^2}{4} = \frac{\pi \times 60^2}{4 \times 100 \times 100} = 0.28 \, \text{m}^2 \tag{7.11}$$

The value of the rod area is calculated as,

$$A_r = \frac{\pi d^2}{4} = \frac{\pi \times 55^2}{4 \times 100 \times 100} = 0.24 \, \text{m}^2 \tag{7.12}$$

The value of the extension force is calculated as,

$$F_e = p \times A_p = 2000 \times 0.28 = 560\,\text{N} \tag{7.13}$$

The value of the retraction force is calculated as,

$$F_r = p \times (A_p - A_r) = 2000 \times (0.28 - 0.24) = 80\,\text{N} \tag{7.14}$$

**Practice Problem 7.1**

A hydraulic system uses a cylinder with a piston diameter of 3.5 in and a rod diameter of 2.5. Calculate the extension force and retraction force of the cylinder if the system's maximum pressure is 2000 psi.

**Practice Problem 7.2**

A hydraulic system uses a cylinder with a piston diameter of 65 mm and a rod diameter of 56 mm. Calculate the extension force and retraction force of the cylinder if the system's maximum pressure is 2500 Pa.

## 7.8 Cylinder Velocity

The cylinder extends with a lower velocity than the retraction time because of the higher piston area. The speeds of the cylinder during extension and retraction are calculated based on the continuity equation. This continuity is expressed as [3],

$$Q = Av \tag{7.15}$$

$$v = \frac{Q}{A} \tag{7.16}$$

In the US customary unit, the unit of the flow rate (Q) is GPM. The conversion of 1 GPM is equal to 231 in$^3$/min and the unit of the area is in$^2$. The extension speed (in/min) is expressed as,

$$v_e = \frac{231Q}{A_p} \tag{7.17}$$

During retraction, the fluid pressure acts on the annular area of the cylinder. Therefore, the speed of the cylinder during retraction is calculated as,

$$v_r = \frac{231Q}{A_p - A_r} \tag{7.18}$$

In the SI unit, 1 LPM is equal to 0.001 m$^3$/min. In this case, the extension speed and the retraction speed are expressed as,

$$v_e = \frac{0.001Q}{A_p} = \frac{Q}{1000A_p} \tag{7.19}$$

$$v_r = \frac{0.001Q}{A_p - A_r} = \frac{Q}{1000(A_p - A_r)} \tag{7.20}$$

*Example 7.3* A 20 GPM pump is used in a hydraulic system to operate a cylinder whose piston diameter is 2.8 in and the rod diameter is 1.5 in. Calculate the extension and retraction speeds of the cylinder.

**Solution**

The value of the piston area is calculated as,

$$A_p = \frac{\pi D^2}{4} = \frac{\pi (2.8)^2}{4} = 6.16\,\text{in}^2 \tag{7.21}$$

The value of the rod area is calculated as,

$$A_r = \frac{\pi d^2}{4} = \frac{\pi (1.5)^2}{4} = 1.77\,\text{in}^2 \tag{7.22}$$

The extension speed is calculated as,

$$v_e = \frac{231Q}{A_p} = \frac{231 \times 20}{6.16} = 750\ \text{in/min} \tag{7.23}$$

The retraction speed is calculated as,

$$v_r = \frac{231Q}{A_p - A_r} = \frac{231 \times 20}{6.16 - 1.77} = 1052.39\,\text{in/min} \tag{7.24}$$

*Example 7.4* A 10 LPM pump is used in a hydraulic system to operate a cylinder whose piston diameter is 60 mm and the rod diameter is 45 mm. Calculate the extension and retraction speeds of the cylinder.

**Solution**

The value of the piston area is calculated as,

$$A_p = \frac{\pi D^2}{4} = \frac{\pi (0.060)^2}{4} = 2.83 \times 10^{-3}\,\text{m}^2 \tag{7.25}$$

The value of the rod area is calculated as,

$$A_r = \frac{\pi d^2}{4} = \frac{\pi (0.045)^2}{4} = 1.59 \times 10^{-3} \, \text{m}^2 \tag{7.26}$$

The extension speed is calculated as,

$$v_e = \frac{Q}{1000 A_p} = \frac{10}{1000 \times 2.83 \times 10^{-3}} = 3.53 \, \text{m/min} = 0.058 \, \text{m/s} \tag{7.27}$$

$$v_r = \frac{Q}{1000(A_p - A_r)} = \frac{10}{1000(2.83 - 1.59) \times 10^{-3}} = 8.06 \, \text{m/min} = 0.13 \, \text{m/s} \tag{7.28}$$

**Practice Problem 7.3**

A 25 GPM pump is used in a hydraulic system to operate a cylinder whose piston diameter is 2.2 in and the rod diameter is 1.1 in. Find the extension and retraction speeds of the cylinder.

**Practice Problem 7.4**

A 15 LPM pump is used in a hydraulic system to operate a cylinder whose piston diameter is 65 mm and the rod diameter is 55 mm. Determine the extension and retraction speeds of the cylinder.

## 7.9  Cylinder Power

The rate of receiving and delivering energy is known as power. Mathematically, the power $P$ is expressed as,

$$P = \frac{dW}{dt} \tag{7.29}$$

In general, Eq. (7.29) is expressed as,

$$P = \frac{W}{t} \tag{7.30}$$

Let us consider a force $F$ is applied to move an object from its initial position to the final position $d$. Then the work done is expressed as,

$$W = F \times d \tag{7.31}$$

Substituting Eq. (7.31) into Eq. (7.30) yields,

$$P = \frac{F \times d}{t} \tag{7.32}$$

However, the velocity v is defined as the distance per unit time and it is expressed as,

$$v = \frac{d}{t} \tag{7.33}$$

Substituting Eq. (7.33) into Eq. (7.32) yields,

$$P = F \times v \tag{7.34}$$

In PFS (Foot, Pound, Second) system, the unit of force is the pound (lb) and the unit of velocity is ft/s. Therefore, the unit of power is lb-ft/s and one horsepower (HP) is equal to 550 lb-ft/s. The power in horsepower is expressed as,

$$P_{HP} = \frac{F \times v}{550} \tag{7.35}$$

In MKS (Meter, Kilogram, Second) system or metric unit or SI unit, the unit of force is Newton (N) and the unit of velocity is m/s. Therefore, the unit of power is N-m/s or W and one kilowatt (HP) is equal to 1000 W. The power in kilowatt is expressed as,

$$P_{kW} = \frac{F \times v}{1000} \tag{7.36}$$

Substituting $p = F/A$ into Eq. (7.34) yields,

$$P = pA \times v \tag{7.37}$$

Again, substituting continuity equation $Q = Av$ into Eq. (7.37) yields,

$$P = pQ \tag{7.38}$$

The unit pressure in the FPS system is lb/in$^2$ (psi) and the unit of flow rate is in$^3$/min. Therefore, the unit of power is lb.in/min. The unit of flow rate in the US customary unit is GPM. The conversion factor, 1 GPM = 231 in$^3$/min is used to convert the flow rate in the required unit. If Q is rated as GPM, then Eq. (7.38) is expressed as,

$$P = p \times 231 Q \quad \frac{\text{lb}}{\text{in}^2} \frac{\text{in}^3}{\text{min}} \tag{7.39}$$

$$P = p \times 231 Q \quad \frac{\text{lb.in}}{\text{min}} \tag{7.40}$$

The following conversion can be written as,

$$1\,\text{HP} = 550 \quad \text{lb.ft/s} \tag{7.41}$$

$$1\,\text{HP} = 550 \times 60 \quad \text{lb.ft/min} \tag{7.42}$$

$$1\,\text{HP} = 550 \times 60 \times 12 \quad \text{lb.in/min} \tag{7.43}$$

Dividing both sides by 396,000 yields,

$$1\text{lb.in/min} = \frac{1}{396000}\,\text{HP} \tag{7.44}$$

Substituting Eq. (7.44) into Eq. (7.40) yields,

$$P_{HP} = \frac{p \times 231Q}{396000} \tag{7.45}$$

$$P_{HP} = \frac{p \times Q}{1714.28} \tag{7.46}$$

Again, the watt is converted to kW as the following,

$$1\,\text{W} = 1\,\text{N m/s} \tag{7.47}$$

$$1\,\text{N m/s} = \frac{1}{1000}\text{kW} \tag{7.48}$$

$$60\,\text{N m/min} = \frac{1}{1000}\text{kW} \tag{7.49}$$

Dividing both sides of Eq. (7.49) by 60 yields,

$$1\,\text{N m/min} = \frac{1}{60 \times 1000}\text{kW} \tag{7.50}$$

In a metric or SI unit, if the power is given as kPa and the flow rate is given as LPM, then Eq. (7.38) is expressed as,

$$P = 1000p\,\left(\text{N/m}^2\right) \times \frac{Q}{1000}\,\left(\text{m}^3/\text{min}\right) = p \times Q \quad \text{N m/min} \tag{7.51}$$

Substituting Eq. (7.50) into Eq. (7.51) yields,

$$P_{kW} = \frac{p \times Q}{60000} \tag{7.52}$$

*Example 7.5*  A 1500 lb load needs to move a certain distance at a speed of 1.5 ft/s. Calculate the required power in HP.

**Solution**

The value of the power in HP is calculated as,

$$P_{HP} = \frac{F \times v}{550} = \frac{1500 \times 1.5}{550} = 4.09 \text{ HP} \qquad (7.53)$$

*Example 7.6*  A 1200 N load needs to move a certain distance at a speed of 1.1 m/s. Determine the required power in kW.

**Solution**

The value of the power in kW is calculated as,

$$P_{kW} = \frac{F \times v}{1000} = \frac{1200 \times 1.1}{1000} = 1.32 \text{ kW} \qquad (7.54)$$

*Example 7.7*  Hydraulic fluid flows through a system with 15 GPM and 1500 psi. Determine the hydraulic power in HP.

**Solution**

The value of the hydraulic power in HP is calculated as,

$$P_{kW} = \frac{p \times Q}{1714.28} = \frac{1500 \times 15}{1714.28} = 13.12 \text{ HP} \qquad (7.55)$$

*Example 7.8*  Hydraulic fluid flows through a system with a 15 LPM and 35,000 kPa. Determine the hydraulic power in kW.

**Solution**

The value of the hydraulic power in kW is calculated as,

$$P_{kW} = \frac{p \times Q}{60000} = \frac{35000 \times 15}{60000} = 8.75 \text{ kW} \qquad (7.56)$$

**Practice Problem 7.5**

A 2500 lb load needs to move a certain distance at a speed of 0.5 ft/s. Calculate the required power in HP.

**Practice Problem 7.6**

An 1800 N load needs to move a certain distance at a speed of 1.2 m/s. Find the required power in kW.

**Practice Problem 7.7**

Hydraulic fluid flows through a system with 25 GPM and 15,000 psi. Calculate the hydraulic power in HP.

**Practice Problem 7.8**

Hydraulic fluid flows through a system with a 35 LPM and 44,000 kPa. Determine the hydraulic power in kW.

## 7.10 Differential Flow of Hydraulic Cylinder

The areas on the cap-end or blind end and rod-end of a double-acting cylinder are different. This difference in areas on the blind and rod end of a double-acting cylinder produces a phenomenon known as differential flow. The differential flow of a hydraulic cylinder is defined as the difference in flow obtained between pushing oil into the blind end while the other end pushes out the oil of the cylinder. The area and the volume at the cap-end and the blind end are not the same as a double-acting cylinder. Let the pump flow rate $Q_p$ be applied at the blind end that pushes the piston with a velocity $v_{Ext}$ and this velocity is considered the same on both sides of the cylinder. At the same time, the $Q_{Return(E)}$ fluid comes out from the rod end and back to the tank as shown in Fig. 7.9.

From continuity equation ($v = Q/A$), the velocity of the cylinder during extension is expressed as,

$$v_{Ext} = \frac{Q_p}{A_p} = \frac{Q_{Return\,(E)}}{A_p - A_r} \tag{7.57}$$

Equation (7.69) is rearranged as,

$$\frac{Q_p}{A_p} = \frac{Q_{Return\,(E)}}{A_p - A_r} \tag{7.58}$$

**Fig. 7.9** Extension of a cylinder

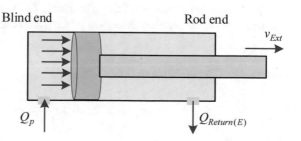

Blind end          Rod end

$v_{Ext}$

$Q_p$        $Q_{Return(E)}$

**Fig. 7.10** Retraction of a
cylinder

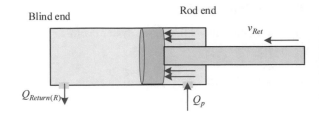

$$Q_{Return\,(E)} = \frac{A_p - A_r}{A_p} Q_p \tag{7.59}$$

where,

$A_p$ is the area of the piston,
$A_r$ is the area of the rod,
$Q_{Ret,E}$ is the amount of fluid return to the tank.

Let the pump flow rate $Q_p$ be applied at the rod end that pushes back the piston with a velocity $v_{Ret}$ as shown in Fig. 7.10. The velocity during retraction is considered the same on both sides of the cylinder.

The amount of fluid $Q_{Return(R)}$ is coming out from the blind end of the cylinder during retraction and return back to the tank. The velocity due to retraction is calculated as,

$$v_{Ret} = \frac{Q_p}{A_p - A_r} = \frac{Q_{Return\,(R)}}{A_p} \tag{7.60}$$

Equation (7.60) is rearranged as,

$$\frac{Q_p}{A_p - A_r} = \frac{Q_{Return\,(R)}}{A_p} \tag{7.61}$$

$$Q_{Return\,(R)} = \frac{A_p}{A_p - A_r} Q_p \tag{7.62}$$

**Example 7.9** A cylinder is used in a system with a pump of a flow rate of 30 GPM. The piston and rod diameters of the cylinder are 2 in and 1.5 in, respectively. Calculate the return flow rates during extension retraction of the cylinder.

**Solution**

The value of the piston area is calculated as,

$$A_p = \frac{\pi D^2}{4} = \frac{\pi (2)^2}{4} = 3.14 \,\text{in}^2 \tag{7.63}$$

The value of the rod area is calculated as,

$$A_r = \frac{\pi D^2}{4} = \frac{\pi (1.5)^2}{4} = 1.77 \, \text{in}^2 \qquad (7.64)$$

The value of the pump return flow during extension is calculated as,

$$Q_{Return\,(E)} = \frac{A_p - A_r}{A_p} Q_p = \frac{3.14 - 1.77}{3.14} \times 30 = 13.09 \, \text{GPM} \qquad (7.65)$$

The value of the pump return flow during retraction is calculated as,

$$Q_{Return\,(R)} = \frac{A_p}{A_p - A_r} Q_p = \frac{3.14}{3.14 - 1.77} \times 30 = 68.76 \, \text{GPM} \qquad (7.66)$$

**Practice Problem 7.9**

A cylinder is used in a system with a pump of a flow rate of 45 LPM. The piston and rod diameters of the cylinder are 55 mm and 45 mm, respectively. Determine the return flow rates during extension retraction of the cylinder.

## 7.11 Cylinder in Horizontal

In this application, a cylinder is placed in a horizontal to push the load as shown in Fig. 7.11. Practically, this cylinder must overcome the friction created between the object and the surface. Let us consider the cylinder that generates the force $F_{Cyl}$ to push the weight of the object is $W$ by overcoming the frictional force or the coefficient of the friction $f$.

Mathematically, it is expressed as,

$$F_{cyl} = f \times W \qquad (7.67)$$

where,

$f$ is dimensionless.

**Fig. 7.11** Cylinder with a horizontal load

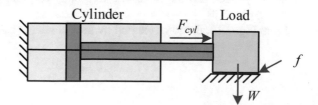

$F_{cyl}$ is the force generated by the cylinder (lb, N),
$W$ is the object weight, (lb, N).

From Eq. (7.57), it is seen that the force exerted by the cylinder can be determined if other parameters are given.

***Example 7.10*** A load of 5000 lb moves by a horizontal cylinder by overcoming the friction factor of 0.11 between the load and the surface. Determine the size of the cylinder if the system pressure is 1000 psi.

**Solution**

The value of the force exerted by the cylinder is calculated as,

$$F_{cyl} = f \times W = 5000 \times 0.11 = 550 \, \text{lb} \tag{7.68}$$

The area of the cylinder is calculated as,

$$A = \frac{F_{cyl}}{p} = \frac{550}{1000} = 0.550 \, \text{in}^2 \tag{7.69}$$

The diameter of the cylinder is calculated as,

$$D = 2\sqrt{\frac{A}{\pi}} = 2\sqrt{\frac{0.550}{\pi}} = 0.84 \, \text{in}^2 \tag{7.70}$$

**Practice Problem 7.10**

A load of 8000 N moves by a horizontal cylinder by overcoming the friction factor of 0.10 between the load and the surface. Find the size of the cylinder if the system pressure is 1000 Pa.

## 7.12  Inclined Cylinder

In this type of application, a cylinder is placed at an angle to lift a load. Let the cylinder be attached with a bearing and placed at an angle θ with the load to lift as shown in Fig. 7.12. The force $F_{bear}$ (bearing force) acts on the bearing when the inclined cylinder lifts the load. The free-body diagram of an inclined cylinder is shown in Fig. 7.13.

From Fig. 7.13, the force exerted by the cylinder is calculated as,

$$\cos \theta = \frac{F_{cyl}}{F_{load}} \tag{7.71}$$

**Fig. 7.12** Inclined cylinder with a load

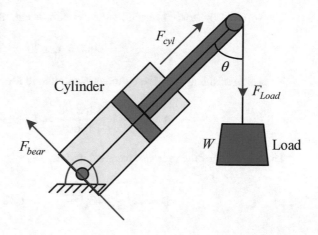

**Fig. 7.13** Free body diagram of an inclined cylinder

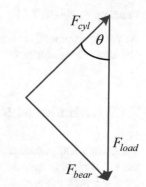

$$F_{cyl} = F_{load} \cos \theta \qquad (7.72)$$

From Fig. 7.13, the force on the bearing is calculated as,

$$\sin \theta = \frac{F_{bear}}{F_{load}} \qquad (7.73)$$

$$F_{bear} = F_{load} \sin \theta \qquad (7.74)$$

***Example 7.11*** A cylinder is placed at an angle of 70° with the horizontal axis to lift a load of 4000 lb. Calculate the diameter of the cylinder if the system pressure is 1000 psi.

**Solution**

The value of the angle with the load is calculated as,

$$\theta = 180° - 90° - 70° = 20° \qquad (7.75)$$

The value of the force exerted by the cylinder is calculated as,

$$F_{cyl} = F_{load} \cos\theta = 4000 \times \cos 20° = 1368.08 \, lb \qquad (7.76)$$

The value of the area of the cylinder is calculated as,

$$A = \frac{F_{cyl}}{p} = \frac{1368.08}{1000} = 1.368 \, in^2 \qquad (7.77)$$

The diameter of the cylinder is calculated as,

$$D = 2\sqrt{\frac{A}{\pi}} = 2\sqrt{\frac{1.368}{\pi}} = 1.32 \, in \qquad (7.78)$$

**Practice Problem 7.11**

A cylinder is placed at an angle of 65° with the horizontal axis to lift a load of 3000 N. Calculate the diameter of the cylinder if the system pressure is 1200 Pa.

## 7.13  Cylinder on Rotating Lever

The common application of a cylinder is to rotate the rotating lever. In this application, the cylinder is attached with a hinge and pin as shown in Fig. 7.14. The cylinder attached with a hinge and pin can exert a maximum force on the lever arm if the angle between the lever arm axis and the cylinder axis is at 90°. Due to this right angle to the lever arm, the lever force turns the lever arm. Let the force due to the cylinder makes an angle θ with the lever arm. The force due to lever can be calculated using the force diagram or free-body diagram as shown in Fig. 7.15. The force due to lever arm is calculated as,

**Fig. 7.14**  Cylinder with a lever arm

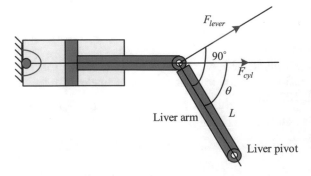

**Fig. 7.15** Free body
diagram of rotating cylinder

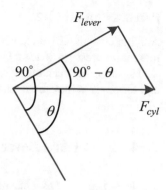

$$\cos(90° - \theta) = \frac{F_{lever}}{F_{cyl}} \tag{7.79}$$

$$\sin \theta = \frac{F_{lever}}{F_{cyl}} \tag{7.80}$$

$$F_{lever} = F_{cyl} \sin \theta \tag{7.81}$$

The torque generated by the lever arm is expressed as the multiplication of the lever force and the distance of the lever arm. Mathematically, it is written as,

$$T = F_{lever} \times L \tag{7.82}$$

Substituting Eq. (7.81) into Eq. (7.82) yields,

$$T = F_{cyl} \sin \theta \times L \tag{7.83}$$

**Example 7.12** A cylinder rotates a lever arm of 15 in length with a maximum pressure of 1000 psi. The diameter of the cylinder piston is 2 in and the angle between the lever arm and the cylinder axis is 40°. Calculate the force and torque generated by the lever arm.

**Solution**

The force exerted by the cylinder is calculated as,

$$F_{cyl} = p \times A = 1000 \times \frac{\pi (2)^2}{4} = 3140 \, lb \tag{7.84}$$

The torque generated by the lever arm is calculated as,

$$T = F_{cyl} \sin \theta \times L = 3140 \times \sin 40° \times 15 = 30275.30 \, lb \text{ - in} \tag{7.85}$$

**Practice Problem 7.12**

A cylinder with a bore diameter of 2.2 in rotates a lever arm of 10 in length with a maximum pressure of 1500 Pa.

The angle between the lever arm and the cylinder axis is 45°. Determine the force and torque generated by the lever arm.

## 7.14   First-Class Lever Arm

The lever is defined as a machine part that can rotate around a fixed point (fulcrum). It can lift and lower the load by the application of less effort. The first-class lever arm is defined as the fulcrum or fixed hinge (pivot) is located in between the cylinder and the load as shown in Fig. 7.16. Examples of the first-class lever arm are hand-pump, beam-balance, punching press hand-wheel etc.

Taking moments around the pivot or fulcrum yields,

$$F_{cyl} \times L_1 - F_{Load} \times L_2 = 0 \tag{7.86}$$

$$F_{cyl} = \frac{L_2}{L_1} \times F_{Load} \tag{7.87}$$

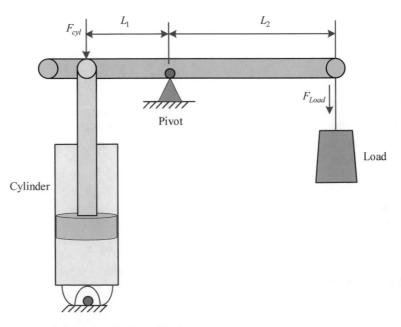

**Fig. 7.16**  Pivot in between cylinder and load

***Example 7.13*** A load of 1500 lb is lifted by a first-class lever arm. The distance from the pivot to the load is 10 in and from the pivot to the load is 7 in. Calculate the force exerted by the cylinder and the size of the cylinder if the system pressure is 1200 psi.

**Solution**

The force exerted by the cylinder is calculated as,

$$F_{cyl} = \frac{L_2}{L_1} \times F_{Load} = 1500 \times \frac{10}{7} = 2142.86 \, \text{lb} \tag{7.88}$$

The size of the cylinder is calculated as,

$$A = \frac{F_{cyl}}{p} = \frac{2142.86}{1200} = 1.78 \, \text{in}^2 \tag{7.89}$$

$$D = 2\sqrt{\frac{A}{\pi}} = 2\sqrt{\frac{1.78}{\pi}} = 1.51 \, \text{in} \tag{7.90}$$

***Example 7.13*** A load of 1400 lb is lifted by a first-class lever arm. The distance from the pivot to the load is 12 in and from the pivot to the load is 5 in. Find the force exerted by the cylinder and the size of the cylinder if the system pressure is 1000 psi.

## 7.15 Second-Class Lever Arm

In the second-class lever arm, the load is attached between the cylinder and the pivot as shown in Fig. 7.17. In this case, the length of the effort-arm (cylinder-arm) is greater than the load-arm from the pivot. The load is attached at a distance of $L_2$ from the pivot and the cylinder is attached at a distance of $L_1$ from the load. A dead-end safety valve is one of the examples.

Taking moments around the pivot or fulcrum yields,

$$F_{cyl} \times (L_1 + L_2) - F_{Load} \times L_2 = 0 \tag{7.91}$$

$$F_{cyl} = \frac{L_2}{L_1 + L_2} \times F_{Load} \tag{7.92}$$

Again, substituting $L = L_1 + L_2$ in Eq. (7.92) yields,

$$F_{cyl} = \frac{L_2}{L} \times F_{Load} \tag{7.93}$$

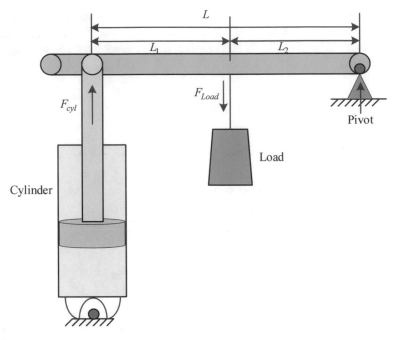

**Fig. 7.17** Pivot at the end of the beam

*Example 7.14* A load of 1500 lb is lifted by a second-class lever arm. The distance from the pivot to the load is 6 in and from the pivot to the cylinder is 21 in. Calculate the force exerted by the cylinder and the size of the cylinder if the system pressure is 900 psi.

**Solution**

The force exerted by the cylinder is calculated as,

$$F_{cyl} = \frac{L_2}{L_1 + L_2} \times F_{Load} = 1500 \times \frac{6}{21 + 6} = 333.33 \, \text{lb} \qquad (7.94)$$

The area and diameter of the cylinder are calculated as,

$$A = \frac{F_{cyl}}{p} = \frac{333.33}{900} = 0.37 \, \text{in}^2 \qquad (7.95)$$

$$D = 2\sqrt{\frac{A}{\pi}} = 2\sqrt{\frac{0.37}{\pi}} = 0.69 \, \text{in} \qquad (7.96)$$

**Practice Problem 7.14**

A load needs to be lifted by a second-class lever arm. The distance from the pivot to the load is 8 in and from the pivot to the cylinder is 16 in. Calculate the load, if the force exerted by the cylinder is 1550 lb.

## 7.16  Third-Class Lever Arm

In the third-class lever arm, the load is attached at one end and the pivot is attached at the other end of the beam. However, the cylinder is attached between the load and pivot as shown in Fig. 7.18.

In this case, the length of the load-arm from the pivot is greater than the effort-arm (cylinder-arm) from the pivot. The load is attached at a distance of $L$ $(L_1 + L_2)$ from the pivot and the cylinder is attached at a distance of $L_2$ from the pivot.

Taking moments around the pivot or fulcrum yields,

$$F_{cyl} \times L_2 - F_{Load} \times (L_1 + L_2) = 0 \tag{7.97}$$

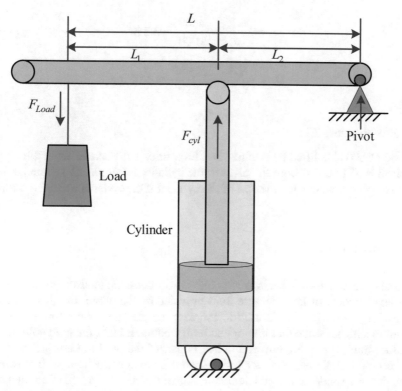

**Fig. 7.18** Pivot at the end of the beam

$$F_{cyl} = \frac{L_1 + L_2}{L_2} \times F_{Load} \tag{7.98}$$

Again, substituting $L = L_1 + L_2$ in Eq. (7.98) yields,

$$F_{cyl} = \frac{L}{L_2} \times F_{Load} \tag{7.99}$$

***Example 7.15*** A load of 900 lb is lifted by a third-class lever arm. The distance from the pivot to the load is 16 in and from the pivot to the cylinder is 6 in. Calculate the force exerted by the cylinder and the size of the cylinder if the system pressure is 1000 psi.

**Solution**

The force exerted by the cylinder is calculated as,

$$F_{cyl} = \frac{L_1 + L_2}{L_2} \times F_{Load} = 900 \times \frac{16 + 6}{6} = 3300 \, \text{lb} \tag{7.100}$$

The area and diameter of the cylinder are calculated as,

$$A = \frac{F_{cyl}}{p} = \frac{3300}{1000} = 3.3 \, \text{in}^2 \tag{7.101}$$

$$D = 2\sqrt{\frac{A}{\pi}} = 2\sqrt{\frac{3.3}{\pi}} = 2.05 \, \text{in} \tag{7.102}$$

**Practice Problem 7.15**

A load of 600 N is lifted by a third-class lever arm. The distance from the pivot to the load is 55 mm and from the pivot to the cylinder is 22 mm. Calculate the force exerted by the cylinder and the size of the cylinder if the system pressure is 1200 Pa.

## 7.17   Intensifier

An intensifier is a device that is used to increase or boost the intensity of the pressure of a large amount of low-pressure fluid provided by the pump. In other words, an intensifier is a device that is used to increase the pressure of fluid in a hydraulic circuit to a higher value that is supplied by the pump. It takes the high volume low-pressure fluid from the pump and converts part of the fluid to high-pressure fluid for accomplishing the necessary task. According to construction, an intensifier is classified as a single-acting and double-acting intensifier. A symbol of an intensifier is shown in Fig. 7.19. Any intensifier has two cylinders such as an input cylinder

**Fig. 7.19** Symbol of an intensifier

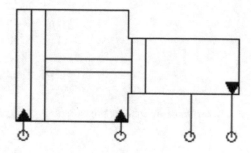

**Fig. 7.20** Intensifier with different pressures and areas

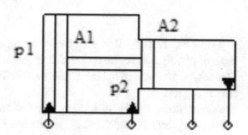

and an output cylinder. The diameter of the input cylinder is higher than the output cylinder. Therefore, the area of the input cylinder is higher than the output cylinder. The pressure and area of each side of the cylinder are shown in Fig. 7.20. The pressure and area of the first cylinder are $p_1$ and $A_1$, respectively. The pressure applied to the input cylinder is expressed as [3, 4],

$$p_1 = \frac{F_1}{A_1} \qquad (7.103)$$

The pressure and area of the second cylinder are $p_2$ and $A_2$, respectively. Then the pressure developed in the second cylinder is expressed as,

$$p_2 = \frac{F_2}{A_2} \qquad (7.104)$$

The input cylinder is physically connected to the output cylinder. Therefore, the force $F_1$ is equal to the force $F_2$. Then Eqs. (7.103) and (7.104) can be rearranged as,

$$p_1 A_1 = p_2 A_2 \qquad (7.105)$$

$$p_2 = \frac{A_1}{A_2} p_1 \qquad (7.106)$$

$$p_2 = \frac{\frac{\pi D_1^2}{4}}{\frac{\pi D_2^2}{4}} p_1 \tag{7.107}$$

$$p_2 = \frac{D_1^2}{D_2^2} p_1 \tag{7.108}$$

In terms of intensifier pressure and pump flow labelling, Eq. (7.106) is modified as,

$$p_{\text{int}} = \frac{A_p}{A_r} p_p \tag{7.109}$$

From Eq. (7.109), it is seen that the output pressure of an intensifier equals the ratio of piston area to the rod area and is multiplied by the input of pump pressure. The output pressure of the intensifier will be higher if the ratio of input area to output area is higher.

Again, consider the piston and the rod moves at the same velocity. The piston velocity ($v_p$) and the rod velocity ($v_r$) are the same and mathematically, it is expressed as,

$$v_p = v_r \tag{7.110}$$

Substituting the continuity equation ($Q = Av$) in Eq. (7.110) yields,

$$\frac{Q_p}{A_p} = \frac{Q_{\text{int}}}{A_r} \tag{7.111}$$

From Eq. (7.111), the ouput flow rate of an intensifier is expressed as,

$$Q_{\text{int}} = \frac{A_r}{A_p} Q_p \tag{7.112}$$

**Example 7.16** An intensifier is used in a 1200 psi and 8 GPM hydraulic system to boost the pressure. The intensifier has a piston area of 12 in² and a rod area of 2 in². Calculate the output pressure and flow rate of an intensifier.

**Solution**

The output pressure of an intensifier is calculated as,

$$p_{\text{int}} = \frac{A_p}{A_r} p_p = \frac{12}{2} \times 1200 = 7200 \, \text{psi} \tag{7.113}$$

The ouput flow rate of an intensifier is calculated as,

$$Q_{int} = \frac{A_r}{A_p}Q_p = \frac{2}{12} \times 8 = 1.33\,\text{GPM} \tag{7.114}$$

***Example 7.17*** An intensifier is used in a 1500 Pa and 20 LPM hydraulic system to boost the pressure. The intensifier has a piston diameter of 60 mm and the rod diameter of 35 mm. Calculate the output pressure and flow rate of an intensifier.

**Solution**

The output pressure of an intensifier is calculated as,

$$p_{int} = \frac{A_p}{A_r}p_p = \frac{60^2}{35^2} \times 1500 = 4408.16\,\text{Pa} \tag{7.115}$$

The ouput flow rate of an intensifier is calculated as,

$$Q_{int} = \frac{A_r}{A_p}Q_p = \frac{35^2}{60^2} \times 20 = 6.81\,\text{LPM} \tag{7.116}$$

**Practice Problem 7.16**

An intensifier is used in a 1500 psi and a flow rate of 10 GPM hydraulic system to boost the pressure. The intensifier has a piston area of 10 in² and a rod area of 2 in². Find the output pressure and flow rate of an intensifier.

**Practice Problem 7.17**

An intensifier is used in a 1400 Pa and a flow rate of 25 LPM hydraulic system to boost the pressure. The intensifier has a piston diameter of 55 mm and the rod diameter of 30 mm. Calculate the output pressure and flow rate of an intensifier.

## 7.18  Application of Intensifier

An intensifier is used in a hydraulic circuit to extend a single-acting cylinder as shown in Fig. 7.21. A fixed displacement pump, pressure relief valve is used to generate fluid flow with lower pressure. A 4/2 lever controlled directional control valve is used to control the fluid flow. Additional fluid is coming from the reservoir and added to the low-pressure fluid. Ultimately, the pressure at the intensifier increase that extends the single-acting cylinder faster.

**Exercise Problems**

7.1  A cylinder with a piston diameter of 2.8 in and a rod diameter of 1.5 in is used in a hydraulic system that has a pressure of 1200 psi. Determine the extension force and retraction force of the cylinder.

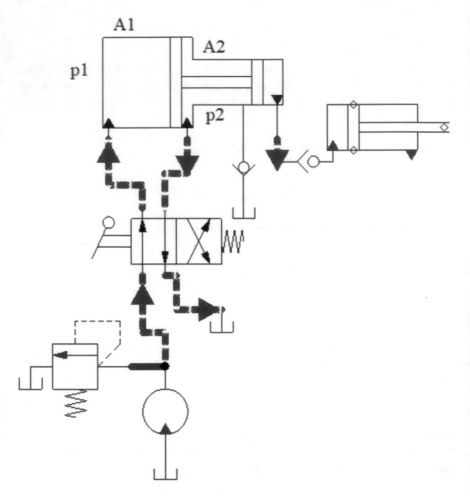

**Fig. 7.21**  Intensifier with a single-acting cylinder

7.2   A hydraulic system uses a cylinder with a piston diameter of 60 mm and a rod diameter of 50 mm. Find the extension force and retraction force of the cylinder if the system's maximum pressure is 1800 Pa.

7.3   A 35 GPM pump is used in a hydraulic system to operate a cylinder whose piston diameter is 3.2 in and the rod diameter is 2 in. Determine the extension and retraction speeds of the cylinder.

7.4   A 25 LPM pump is used in a hydraulic system to operate a cylinder whose piston diameter is 55 mm and the rod diameter is 35 mm. Calculate the extension and retraction speeds of the cylinder.

7.5   A 2100 lb load needs to move a certain distance at a speed of 0.5 ft/s. Calculate the necessary power in HP.

7.6    An 2300 N load needs to move a certain distance at a speed of 3.5 m/s. Find the required power in kW.

7.7    Hydraulic fluid flows through a system with 35 GPM and 11,000 psi. Calculate the hydraulic power in HP.

7.8    Hydraulic fluid flows through a system with a 25 LPM and 25,000 kPa. Determine the hydraulic power in kW.

7.9    A cylinder is used in a system with a pump of a flow rate of 35 GPM. The piston and rod diameters of the cylinder are 2.5 in and 1.4 in, respectively. Calculate the return flow rates during extension retraction of the cylinder.

7.10   A cylinder is used in a system with a pump of a flow rate of 25 LPM. The piston and rod diameters of the cylinder are 60 mm and 50 mm, respectively. Determine the return flow rates during extension retraction of the cylinder.

7.11   A load of 7000 lb moves by a horizontal cylinder by overcoming the friction factor of 0.12 between the load and the surface. Find the size of the cylinder if the system pressure is 1200 psi.

7.12   A load of 8500 N moves by a horizontal cylinder by overcoming the friction factor of 0.10 between the load and the surface. Find the size of the cylinder if the system pressure is 1200 Pa.

7.13   A cylinder is placed at an angle of 35° with the horizontal axis to lift a load of 3500 N. Calculate the diameter of the cylinder if the system pressure is 1200 Pa.

7.14   A cylinder rotates a lever arm of 12 in length with a maximum pressure of 1200 psi. The diameter of the cylinder piston is 2.2 in and the angle between the lever arm and the cylinder axis is 30°. Calculate the force and torque generated by the lever arm.

7.15   A load of 1500 lb is lifted by a first-class lever arm. The distance from the pivot to the load is 10 in and from the pivot to the load is 6 in. Find the force exerted by the cylinder and the size of the cylinder if the system pressure is 1200 psi.

7.16   A load needs to be lifted by a second-class lever arm. The distance from the pivot to the load is 8 in and from the pivot to the cylinder is 15 in. Calculate the load, if the force exerted by the cylinder is 1240 lb.

7.17   A load of 1000 lb is lifted by a third-class lever arm. The distance from the pivot to the load is 18 in and from the pivot to the cylinder is 7 in. Calculate the force exerted by the cylinder and the size of the cylinder if the system pressure is 1200 psi.

7.18   A load of 1600 N is lifted by a third-class lever arm. The distance from the pivot to the load is 60 mm and from the pivot to the cylinder is 30 mm. Calculate the force exerted by the cylinder and the size of the cylinder if the system pressure is 1400 Pa.

7.19   An intensifier is used in a 1000 psi and 10 GPM hydraulic system to boost the pressure. The intensifier has a piston area of 15 in$^2$ and a rod area of 3 in$^2$. Calculate the output pressure and flow rate of an intensifier.

7.20    An intensifier is used in a 1400 Pa and 25 LPM hydraulic system to boost the
        pressure. The intensifier has a piston diameter of 55 mm and the rod diameter
        of 40 mm. Calculate the output pressure and flow rate of an intensifier.


# References

1. A. Esposito, *Fluid Power with Applications*, 7th edn. (Pearson New International Education,
   USA, 2014), pp. 1–648
2. J.L. Johnson, *Introduction to Fluid Power*, 1st edn. (Delmar, a division of Thomson Learning,
   USA, 2002), pp. 1–502
3. F. Don Norvelle, *Fluid Power Technology*, 1st edn. (Delmar, a division of Thomson Learning,
   USA, 1996), pp. 1–649
4. J.A. Sullivan, *Fluid Power-Theory And Applications*, 3rd edn. (Prentice-Hall New Jersey, USA,
   1989), pp. 1–528

# Chapter 8
# Hydraulic Pressure Control Devices

## 8.1 Introduction

Pressure control in a hydraulic system is important to control the force of a cylinder and the torque of the motor. However, the force is of a cylinder equals the product of the system fluid pressure and the area over which the pressure applied. The pressure control devices or valves are used to determine and sclcct the pressure levels of certain machines to operate safely. The pressure in the hydraulic system rises due to the pump flow rate being higher than the flow rate through the cylinder, the volume of the closed system is reduced, etc. The pressure controls are used to limit the system pressure at a safe level, reducing the pressure based on the machine rating, unloading the pressure, and helping the sequencing operation of cylinders. The pressure control tasks are accomplished by the group of valves. This group includes pressure relief valve, pressure reducing valve, pressure sequence valve, unloading valve, loading valves, and brake valve. Details of those valves will be discussed in this chapter.

## 8.2 Pressure Relief Valves

The pressure relief valves are used to maintain the system maximum pressure. This type of task is accomplished by converting the part of the fluid flow to the tank or reservoir when the pressure reaches a preset value. A direct-type pressure relief valve is shown in Fig. 8.1, a ball or poppet is on one side and the spring on the other side depending on the pump pressure. A poppet is held seated inside the valve by a heavy spring.

When the pressure in the pump creates the force on the ball that is less than the spring pressure, then the ball remains on the seat and all the pump flow will pass to the hydraulic circuit or system. When the pressure in the pump creates a force on the ball that is higher than the spring pressure, the ball will move off its seat. This condition will allow the pump flow to go back to the tank through the pressure

© The Author(s), under exclusive license to Springer Nature Singapore Pte Ltd. 2022     287
Md. A. Salam, *Fundamentals of Pneumatics and Hydraulics*,
https://doi.org/10.1007/978-981-19-0855-2_8

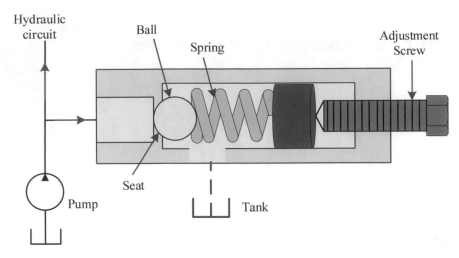

**Fig. 8.1** Schematics of a pressure relief valve

relief valve as long as the system pressure maintains the maximum level. The value of the pressure at which the valve just begins to open and bypass the fluid to the tank is known as cracking pressure. At this pressure, the force developed by the fluid on the ball or poppet to overcome the compressive force by the spring retaining the ball on the seat. The pressure at which the pressure relief valve is fully open and full flow of the pump bypasses the fluid to the tank is known as full-flow pressure. The difference between the full flow pressure and the cracking pressure is known as pressure override. The pressure relief valve is similar to a circuit breaker in an electrical circuit.

The full pump flow will return to the tank through the pressure relief valve if the hydraulic system does not receive or accept any flow. The pressure relief valve also protects against the overload that is experienced by the cylinders and associated components in the hydraulic system. Another important function of the pressure relief valve is to limit the force or torque produced by hydraulic cylinders or motors.

The graphic symbol of a pressure relief valve is shown in Fig. 8.2. The pressure line of the graphic symbol works against the spring pressure and opens the pressure relief valve when the force of the pump flow rate is higher than the spring compression

**Fig. 8.2** Graphic symbol of a pressure relief valve

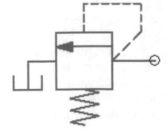

pressure. According to the graphic symbol of the pressure relief valve, the valve is normally closed and on one side of the valve, pressure is fed (dotted lines) to open it. Whereas the spring tries to provide an adjustable pressure level at which the relief valve opens.

The power in the hydraulic circuit is discussed in the previous chapters. Mathematically, the power loss across the pressure relief valve is calculated by the following equation [2],

$$P_{kW} = \frac{pQ}{1000} \tag{8.1}$$

where,

$p$ is the pressure in N/m$^2$,
$Q$ is the flow rate in m$^3$/s.

The power loss across pressure relief can also be calculated as,

$$P_{hp} = \frac{pQ}{1714.28} \tag{8.2}$$

where,

$p$ is the pressure in psi,
$Q$ is the flow rate in GPM.

The force due to the valve fully closed ($F_{vc}$) from the initial position is calculated as the product of spring constant and the initial displacement of the valve as,

$$F_{vc} = kS_{initial} \tag{8.3}$$

The cracking pressure ($p_{cp}$) is calculated as the ratio of force ($F_{vc}$) and the area of the poppet ($A_{poppet}$) as,

$$p_{cp} = \frac{F_{vc}}{A_{poppet}} \tag{8.4}$$

The force required to fully open the valve ($F_{fo}$) is calculated as the product of spring constant and the final displacement ($S_{fd}$) of the valve as,

$$F_{fo} = kS_{fd} \tag{8.5}$$

The fully pump pressure ($p_{fpp}$) is calculated as the ratio of force ($F_{fo}$) and the area of the poppet ($A_{poppet}$) as,

$$p_{fpp} = \frac{F_{fo}}{A_{poppet}} \qquad (8.6)$$

where the unit of spring constant is N/m or lb/ft.

***Example 8.1*** The pressure setting of a pressure relief valve is 100 bar. It returns all the flow from a 0.0008 m$^3$/s pump to the tank. Calculate the power loss across the valve.

**Solution**

The value of the piston area is calculated as,

$$P_{kW} = \frac{pQ}{1000} = 100 \times 10^5 \times 0.0008 = 8\,\text{kW} \qquad (8.7)$$

**Practice Problem 8.1**

The power loss across the pressure relief valve is found to be 20 kW. It returns all the flow from a 0.00012 m$^3$/s pump to the tank. Calculate the pressure setting of a pressure relief valve.

***Example 8.2*** The pressure setting of a pressure relief valve is 1200 psi. It returns all the flow to the tank from a 25 GPM pump. Determine the power loss across the valve.

**Solution**

The power loss across pressure relief can also be calculated as,

$$P_{hp} = \frac{pQ}{1714.28} = \frac{1200 \times 25}{1714.28} = 17.5\,\text{HP} \qquad (8.8)$$

**Practice Problem 8.2**

A pressure relief valve returns all the flow to the tank from a 15 GPM pump. If the power loss across the valve is 20 HP, find the pressure setting of the pressure relief valve.

***Example 8.3*** The area of the poppet of a pressure relief valve is found 3 cm$^2$. A spring constant is set to 2500 N/cm to hold the poppet at the seat. The adjustment knob is then set to 0.35 cm initially from its free length condition. The valve is used to pass full pump flow to the tank at the valve pressure setting, the poppet must be moved 0.25 cm from its fully closed position. Determine the cracking pressure and the full pump flow pressure.

**Solution**

The force due to fully closed the valve from the initial displacement is calculated as,

$$F_{vc} = kS_{initial} = 2500 \times 0.35 = 875\,N \tag{8.9}$$

The cracking pressure is calculated as,

$$p_{cp} = \frac{F_{vc}}{A_{poppet}} = \frac{875}{3 \times 10^{-4}} = 2916.67\,kPA \tag{8.10}$$

The force due to fully open is calculated as,

$$F_{fo} = kS_{final} = 2500 \times 0.(35 + 0.25) = 1500\,N \tag{8.11}$$

The full pump flow pressure is calculated as,

$$p_{fpp} = \frac{F_{fo}}{A_{poppet}} = \frac{1500}{3 \times 10^{-4}} = 5000\,kPA \tag{8.12}$$

**Practice Problem 8.3**

The area of the poppet of a pressure relief valve is found 3.14 in². A spring constant is set to 1500 lb/in to hold the poppet at the seat. The adjustment knob is then set to 0.25 in initially from its free length condition. The valve is used to pass full pump flow to the tank at the valve pressure setting, the poppet must be moved 0.12 in from its fully closed position. Find the cracking pressure and the full pump flow pressure.

# 8.3  Pressure Reducing Valves

The second type of pressure control family is a pressure-reducing valve. The pressure reducing valves are used in a fluid power to operate one or more branches at pressures lower than the system maximum pressure. In a brief, these types of valves are used in a hydraulic circuit where the reduced pressure needs to maintain. The pressure-reducing valve is normally open [3].

It is actuated by the system downstream pressure and tends to close as the pressure reaches the valve setting pressure. The pressure-reducing valve uses a spring-loaded spool to control the downstream pressure. If the pressure at the output is less than the spring setting pressure, then the spool moves to the left allowing free flow from the input line to the output line as shown in Fig. 8.3. The spool moves to the right

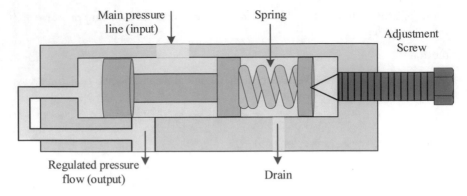

**Fig. 8.3** Schematics of a pressure reducing valve

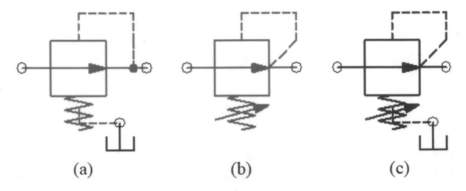

**Fig. 8.4** Schematics of a pressure reducing valve

to block the output line port as the pressure increases at the output line. The graphic symbols of a pressure reducing valve with drain, variable pressure reducing valve with and without drains are shown in Fig. 8.4. An external drain is necessary on a pressure reducing valve because both sides of this valve are under pressure.

## 8.4   Application Pressure Relief and Reducing Valves

The application of pressure relief valve and pressure reducing is shown in Fig. 8.5. In this application, two double-acting cylinders are connected in parallel. The first cylinder in this circuit operates at the maximum pressure which is determined by the pressure relief valve. The second cylinder is operated at a lower pressure. This lower pressure is achieved by placing a pressure reducing valve as shown in Fig. 8.5.

For any reason, if the pressure in the second cylinder rises above the pre-setting operating pressure, the pressure reducing valve closes partially to create a pressure

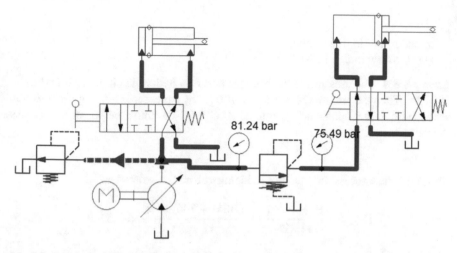

**Fig. 8.5** Application of pressure reducing valve

drop across the valve. Therefore, the pressure reducing valve then maintains the pressure drop so that the output pressure is not allowed to rise above the pre-setting setting. The pressure drop across the reducing valve represents the lost energy that is being converted into heat that is the main disadvantage of this method. In any case, if the pressure setting of the reducing valve is set to very low relative to the pressure in the rest of the system, the pressure drop will be very high, resulting in excessive heating of the fluid. As a result, the hydraulic oil becomes too hot which in turn reduces viscosity. Ultimately, this situation will increase component wear and tear. Let the pressure at the primary circuit be $p_1$ and the pressure at the secondary circuit is $p_2$. Then the power loss across the pressure reducing valve can be calculated in the SI unit as,

$$P_{loss} = \frac{(p_1 - p_2)Q}{60000} \tag{8.13}$$

where,

$p_1$ and $p_2$ are in N/m$^2$ (Pa),
$Q$ is the flow rate in GPM.

In the FPS system, it is calculated as,

$$P_{loss} = \frac{(p_1 - p_2)Q}{1714.28} \tag{8.14}$$

where,

$p_1$ and $p_2$ are in lb/in$^2$ (psi),
$Q$ is the flow rate in LPM.

**Example 8.4** The pressure at the primary part of a hydraulic circuit is found to be 1000 psi. Then the pressure is reduced to 700 psi by a pressure reducing valve for the second circuit at a constant flow rate of 20 GPM. Determine the power loss across the valve.

**Solution**

The power loss across the pressure reducing valve is calculated as,

$$P_{loss} = \frac{(p_1 - p_2)Q}{1714.28} = \frac{(1000 - 700) \times 20}{1714.28} = 3.5\,\text{HP} \qquad (8.15)$$

**Example 8.5** The pressure at the primary part of a hydraulic circuit is found to be 1200 Pa. Then the pressure is reduced to 900 Pa by a pressure reducing valve for the second circuit at a constant flow rate of 25 LPM. Calculate the power loss across the valve.

**Solution**

The power loss across the pressure reducing valve is calculated as,

$$P_{loss} = \frac{(p_1 - p_2)Q}{60000} = \frac{(1200 - 900) \times 25}{60000} = 0.125\,\text{kW} \qquad (8.16)$$

**Practice Problem 8.4**

The pressure at the primary part of a hydraulic circuit is found to be 1500 psi. Then the pressure is reduced to 900 psi by a pressure reducing valve for the second circuit at a constant flow rate. Determine the flow rate if the power loss across the valve is 5 HP.

**Practice Problem 8.5**

The pressure at the primary part of a hydraulic circuit is found to be 1400 Pa. Then the pressure is reduced to a certain value by a pressure reducing valve for the second circuit at a constant flow rate of 35 LPM. Determine the pressure at the second circuit if the power loss across the valve is 2 kW.

## 8.5 Pressure Sequence Valve

The sequence valve is used to control the operation of cylinders sequentially. The cylinder in the sequence with the least resistance runs first smoothly with a single

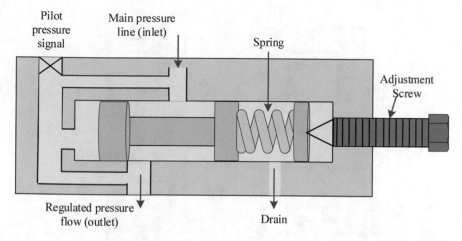

Pilot pressure signal

Main pressure line (inlet)

Spring

Adjustment Screw

Regulated pressure flow (outlet)

Drain

**Fig. 8.6**  Schematic of a sequence valve

directional control valve. Whereas the cylinder in the sequence with the highest resistance needs an additional directional control valve to maintain the sequence operation. There is another way to force the fluid to move through the highest resistance path by a sequence valve. A sequence valve is a normally closed poppet valve that opens at an adjustable set pressure as shown in Fig. 8.6. It has an external drain port to keep the leaking fluid. This valve also uses a check valve to bypass the fluid from the output to the input.

The graphic symbols of sequence valves are shown in Fig. 8.8.

The graphic symbols of a sequence valve and the variable sequence valve are shown in Fig. 8.7a, b. Whereas the graphic symbols of a sequence valve with the check and a variable sequence valve with the check are shown in Fig. 8.7c, d. The graphic symbols of a remote sequence valve and a remote variable sequence valve are shown in Fig. 8.7e, f. The physical sequence valve is shown in Fig. 8.8.

## 8.6  Application of Sequence Valve

The horizontal and the vertical cylinders are connected in parallel as shown in Fig. 8.9. The horizontal cylinder extends first to place and hold the working piece at the required position by pressing the lever of the 4/3 DCV. After extension, the sequence valve works and extends the vertical cylinder for shaping the workpiece as shown in Fig. 8.10. Then the lever is pressed and both the cylinders are retracted at the same time as shown in Fig. 8.11.

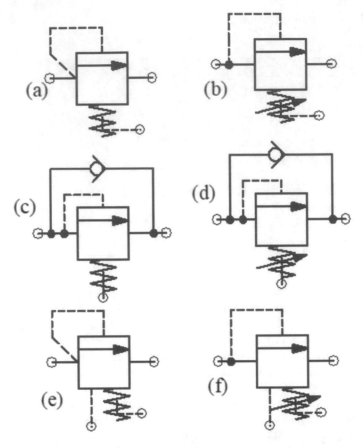

**Fig. 8.7** Symbols of sequence valves

**Fig. 8.8** Physical sequence valve

**Fig. 8.9** Horizontal and
vertical cylinders

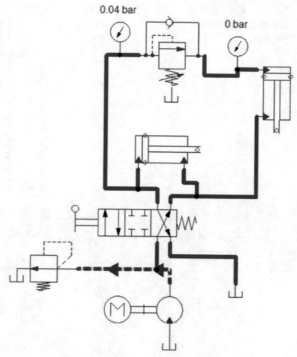

**Fig. 8.10** Extension of
horizontal and vertical
cylinders

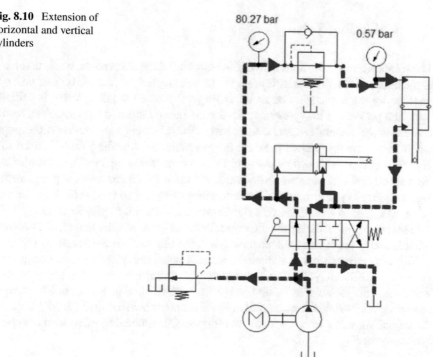

**Fig. 8.11** Retraction of horizontal and vertical cylinders

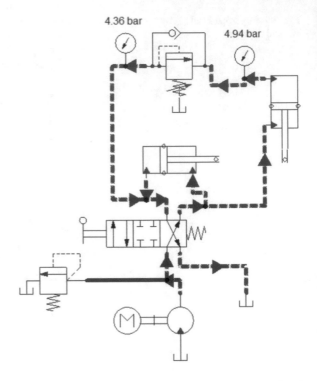

## 8.7　Unloading Valve

There are certain applications in hydraulic circuits where the pressure needs to release to the tank instead of passing through the pressure relief valve. This type of task can be done by the application of an unloading valve. An unloading valve is initially open and provides a low-pressure path for the pump output to the tank. As a result, it reduces the system's pressure near zero. An unloading valve unloads the pump when the maximum system pressure is achieved. An unloading valve is used in a high flow pump and low flow pump circuits where two pumps move a cylinder at a high speed and low pressure. Afterward, the circuit shifts to a single pump with a high pressure to perform work. The unloading valve is also used at the cap-end side of an oversized to send the excess flow to the tank when the cylinder retracts.

This arrangement makes the directional control valve smaller in size. This valve is reliable and requires less maintenance. An unloading valve is used in many places to perform operations such as stamping, coining, punching, piercing, etc. The graphic symbols of different types of unloading valves are shown in Fig. 8.12. The unloading valve and the variable unloading valve are shown in Fig. 8.12a, b. The remote unloading and variable remote unloading valves are shown in Fig. 8.12c, d. Similarly, the unloading valve with the check and the variable unloading valve with the check are shown in Fig. 8.12c, f.

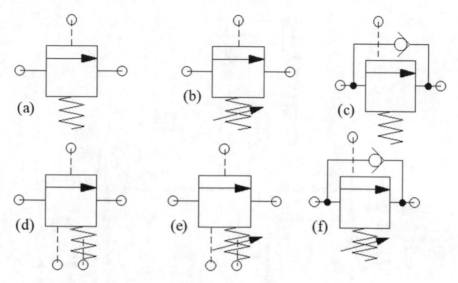

**Fig. 8.12**  Symbols of unloading valves

## 8.8  Application of Unloading Valve

The unloading valve, pressure relief valve, cylinder, and check valves are connected as shown in Fig. 8.13. The high flow pump and low flow pump will supply pressure when the system pressure is less than 1000 psi. When the pump flow reaches 1000 psi, then the unloading valve opens and unloads or releases the flow from a high flow pump to the tank at low pressure as shown in Fig. 8.14.

When the system pressure is less than 1000 psi, then both the pumps work and supply to the system. As a result, the cylinder will extend if the lever of the DCV is pushed as shown in Fig. 8.15.

The cylinder will retract when the lever of the DCV is pulled as shown in Fig. 8.16.

## 8.9  Counterbalance Valve

The hydraulic cylinders are used in different applications such as moving a horizontal load, raising and dropping the loads vertically and inlined mounted. An accident can happen if anything goes wrong when lifting and dropping heavy loads. Therefore, counterbalance valves are also known as holding valves are used to prevent loads from dropping quickly or uncontrollably. This type of load holding capacity of a counterbalance valve is achieved by allowing free flow into the cylinder and by preventing the reverse flow until a pilot pressure inversely proportional to the load is applied. In brief, the counterbalance valve permits flow in one direction and blocks the flow in opposite direction. Counterbalance valves are used in many applications

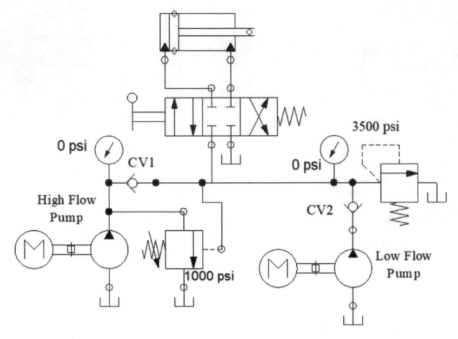

**Fig. 8.13** High and low flow pump with an unloading valve

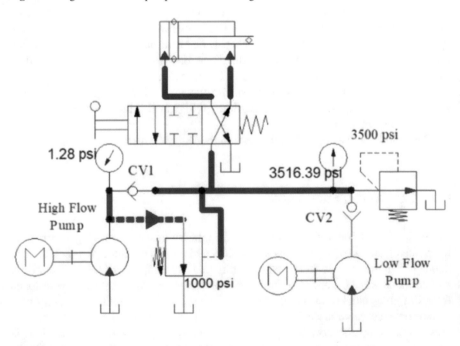

**Fig. 8.14** Both unloading and pressure relief valves open

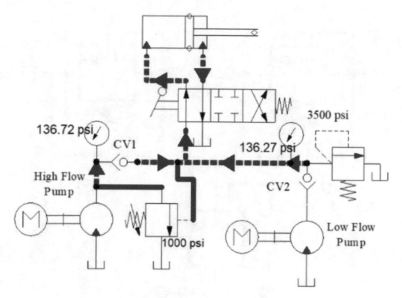

**Fig. 8.15** Both pumps supply during extension

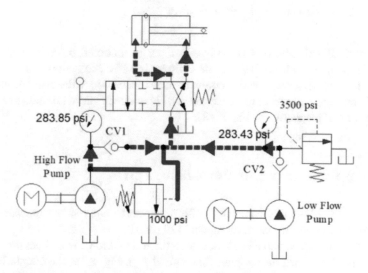

**Fig. 8.16** Both pumps supply during retraction

such as vertical presses, forklifts, loaders, etc. There are many advantages of the counterbalance valve are to prevent uncontrolled movement of both the static or dynamic loads, protect the associated equipment from damages by induced pressure, control the load speed in case of hydraulic hose failure and stabilize the load when the system experiences high or variable backpressure. It is also used for clamping

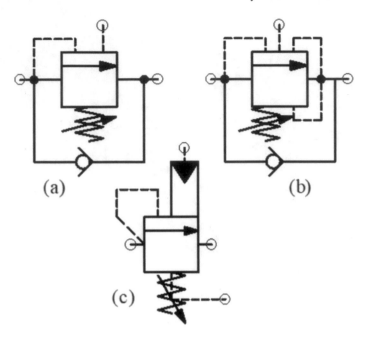

**Fig. 8.17** Graphic symbols of counterbalance valves

applications. Practically, this valve is either flange mounted or connected with a metallic pipe and is placed next to the cylinder to offer protection. The graphical symbols for counterbalance valve, counterbalance valve with a drain not affected by the backpressure, and counterbalance valve without check and with drain are shown in Fig. 8.17a–c, respectively (Fig. 8.18).

## 8.10  Application of Counterbalance Valve

The hydraulic cylinder is connected with a counterbalance valve, pressure relief valve, directional control valve as shown in Fig. 8.19.

The counterbalance valve is connected at the rod end to provide backpressure to stabilize the downward movement of the cylinder. In Fig. 8.19, the counterbalance valve is operated remotely as the pilot line is connected with the cap end line of the cylinder. The counterbalance valve has the check valve to bypass the hydraulic fluid when the cylinder is retracted. The pressure drop across the counterbalance valve is converted to heat that is the main disadvantage of this valve. The cylinder is extended by pressing the lever of the DCV and the fluid is passing through the counterbalance valve as shown in Fig. 8.20.

Similarly, the cylinder is retracted by pulling the lever of the DCV. In this case, the fluid is passing through the check valve as shown in Fig. 8.21.

**Fig. 8.18** Graphic symbols
of counterbalance valves

**Fig. 8.19** Connection of
counterbalance valve with a
cylinder

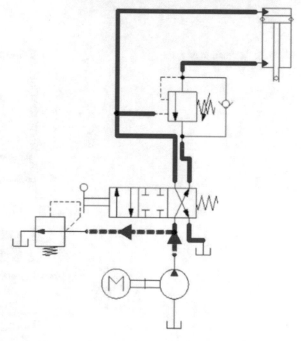

**Fig. 8.20** Extension of a
cylinder

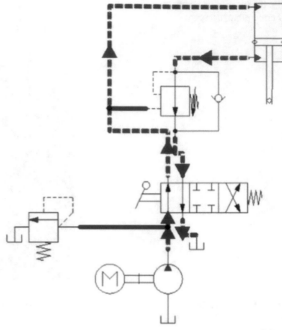

**Fig. 8.21** Retraction of a
cylinder

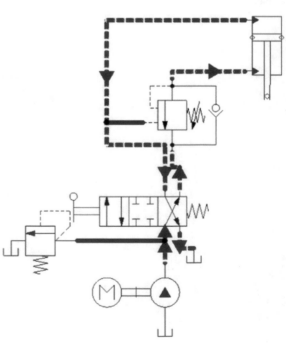

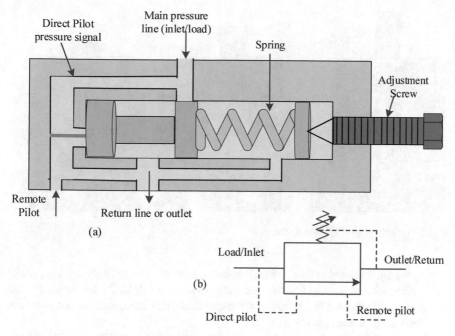

**Fig. 8.22** Brake valve and graphical symbol

## 8.11 Brake Valve

The brake valve, like a counterbalance valve, prevents the load from abruptly or forcefully accelerating. The brake valve is used in a hydraulic circuit with the hydraulic motor to reduce the weight of the various sizes. In some cases, the load may drive the motor rather than the motor driving the large load down. This condition is known as overrunning the load. The brake valve is not available in the Automation Studio software. However, the main schematic of a brake valve and the graphical symbol is shown in Fig. 8.22.

From Fig. 8.22b, it is seen that the brake valve is normally closed and it can be opened either by direct piloting pressure or remote piloting pressure.

## 8.12 Flow Control Valve

An orifice is also known as a flow restrictor is used in fluid power to control air, gas, and hydraulic fluid. An orifice has a fixed or variable diameter to control the fluid flow. Flow control is obtained by opening and closing the diameter of the valve (Fig. 8.23).

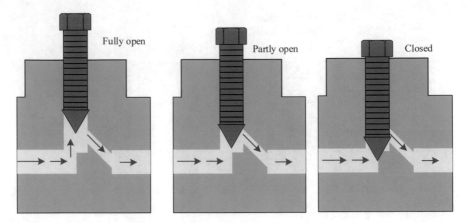

**Fig. 8.23** Orifice with different positions

In general, it is used to reduce the flow, increase fluid velocity, and precisely meter flow. Manufacturer of fluid power components such as SMC, Parker Festo printed the graphic symbol on the surface of components. The input and output ports are decided based on the application.

The one-way non-return throttle valve is made by the combination of the check valve and an orifice. The check valve allows the fluid to flow in one direction, whereas it blocks the fluid flow in the opposite direction. The check valve is connected in parallel with an orifice with a variable diameter to form a one-way non-return throttle valve or simply a flow control valve. The graphic symbol of an orifice and the flow control valve is shown in Fig. 8.24.

The physical flow control valves are shown in Figs. 8.25 and 8.26. These valves are used in a pneumatic trainer kit and hydraulic trainer kit to control the flow rate.

**Fig. 8.24**  Graphic symbols of an orifice and flow control valve

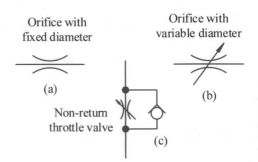

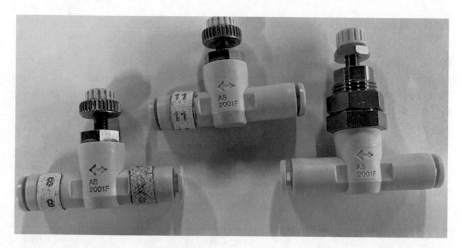

**Fig. 8.25** Physical pneumatic flow control valve

**Fig. 8.26** Physical hydraulic flow control valve

## 8.13 Flow Divider

A flow divider is a fluid power component that is used to divide the flow equally with the same flow rates. There are two types of flow divider such as spool-type and rotary type are normally used in a hydraulic circuit. The graphical symbol and physical flow divider are shown in Fig. 8.27.

**Fig. 8.27** Graphical symbol and physical flow divider

## 8.14   Orifice Equation

Let us consider a tank full of liquid and an orifice is placed at the bottom of the tank as shown in Fig. 8.28. The diameter of the orifice at the initial position is $D_0$ and then the diameter is reduced known as contraction. It is also known as vena contracta and the diameter of the vena contracta or contraction is $D_c$. Applying Bernoulli's equation in a pressure head form yields,

$$\frac{p_1}{\rho g} + \frac{v_1^2}{2g} + z_1 = \frac{p_2}{\rho g} + \frac{v_2^2}{2g} + z_2 \qquad (8.17)$$

From Fig. 8.28, the following equations are written as,

$$\frac{p_1}{\rho g} = H \qquad (8.18)$$

$$\frac{p_2}{\rho g} = 0 \text{ (At atmospheric pressure)} \qquad (8.19)$$

$$z_1 = z_2 \qquad (8.20)$$

The velocity $v_1 \ll v_2$, then $v_1 = 0$. Substituting equations from (8.18) to (8.20) into Eq. (8.17) yields,

$$H = \frac{v_2^2}{2g} \qquad (8.21)$$

$$v_2 = \sqrt{2gH} \qquad (8.22)$$

Equation (8.22) is known as Torricelli's equation.

The ratio of actual velocity ($v_a$) to the theoretical velocity ($v_t$) is known as velocity coefficient ($C_v$) and mathematically, it is expressed as,

**Fig. 8.28**  A tank with an orifice

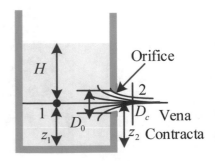

$$C_v = \frac{v_a}{v_t} \tag{8.23}$$

Substituting Eq. (8.22) into Eq. (8.23) yields,

$$v_{2a} = C_v\sqrt{2gH} \tag{8.24}$$

The contraction coefficient (Cc) is defined as the ratio of output area ($A_c$) to the input area ($A_0$) and mathematically, it is expressed as,

$$C_c = \frac{A_2}{A_0} \tag{8.25}$$

$$A_2 = C_c A_0 \tag{8.26}$$

According to the continuity equation, the actual flow rate at point 2 is,

$$Q_{2a} = A_2 v_{2a} \tag{8.27}$$

Substituting Eqs. (8.24) and (8.26) into Eq. (8.27) yields,

$$Q_{2a} = C_c A_0 C_v \sqrt{2gH} \tag{8.28}$$

The discharge coefficient (Cd) is defined as the ratio of actual discharge to theoretical discharge. Mathematically, it is expressed as,

$$C_d = \frac{Q_a}{Q_t} = C_c C_v \tag{8.29}$$

Substituting Eq. (8.29) into Eq. (8.28) yields,

$$Q_a = C_d A_0 \sqrt{2gH} \tag{8.30}$$

The typical value of $C_c$ and $C_v$ for a sharp-edged orifice is 0.66 and 0.98, respectively.

## 8.15 Flow Coefficient

The flow rate is already defined in the relevant chapter. However, it is defined as the volume amount of fluid passing through a pipe per unit of time. The flow rate through any valve is proportional to the pressure drop across it. The pressure drop will be large for a higher flow rate through a valve. Therefore, the relationship between the flow rate and the pressure drop. The flow coefficient is an important measurement

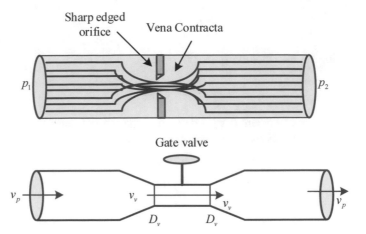

**Fig. 8.29** Pipe with a sharp-edged orifice

that helps to calculate how much fluid can pass through a valve. Therefore, the flow coefficient ($C_v$) can measure the rate at which any gas or liquid can pass through a valve. There are two ways to calculate the pressure drop across the valve or pipe. One of the methods is applying Bernoulli's equation. Applying Bernoulli's equation in Fig. 8.29 to calculate the head loss of the valve and the pipe.

The head loss across the pipe is expressed as,

$$h_L = \frac{\Delta p}{\rho g} = \frac{\Delta p}{\gamma} \qquad (8.31)$$

The head loss across the gate valve is calculated as,

$$h_L = k_v \frac{v_v^2}{2g} \qquad (8.32)$$

where $v_v$ is the velocity of the fluid inside the valve.

From Eqs. (8.31) and (8.32), the following equation is written as,

$$h_L = \frac{\Delta p}{\gamma} = k_v \frac{v_v^2}{2g} \qquad (8.33)$$

The valve of valve coefficient ($k_v$) increases as the valve opening decreases. The characteristics of fluid flow inside the valve are shown in Fig. 8.30. At a certain point, the pressure inside the vena contracta will reach the vapour pressure.

However, the flow rate inside the gate valve is linearly proportional to the square root of the pressure difference between the upstream and downstream. Mathematically, it is expressed as,

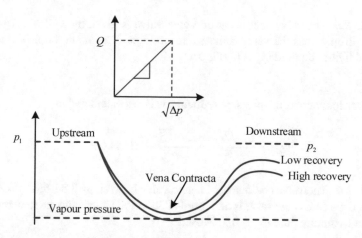

**Fig. 8.30** Characteristics of fluid inside the valve

$$Q = C_v \sqrt{\Delta p} \qquad (8.34)$$

Where Cv is valve flow coefficient. The Cv is the discharge (US customary unit) of water at 60°F that passes through a valve with a pressure difference of 1 psi. When a liquid is passing through a valve, the specific gravity of the fluid needs to be considered and the flow rate is inversely proportional to the square root of the specific gravity. Mathematically, it is expressed as [4],

$$Q = \frac{1}{\sqrt{S_g}} \qquad (8.35)$$

Combining Eqs. (8.34) and (8.35) yields the general design equation as,

$$Q = \frac{C_v \sqrt{\Delta p}}{\sqrt{S_g}} \qquad (8.36)$$

From Eq. (8.36), the pressure drop is expressed as,

$$\Delta p = \frac{S_g Q^2}{C_v^2} \qquad (8.37)$$

where,

$Q$ is the flow rate in GPM, LPM,
$\Delta p$ is the pressure drop across the valve in psi, kPa,
$S_g$ is the specific gravity in dimensionless,
$C_v$ is the flow coefficient in GPM/$\sqrt{}$(psi), LPM/$\sqrt{}$(kPa).

*Example 8.6* The valve coefficient of a gate valve is given by 2 GPM/$\sqrt{}$(psi). The pressure drop across the valve is measured at 25 psi. Calculate the flow rate if the saturated hydraulic fluid ($S_g$ is 0.9) is used.

**Solution**

The power loss across the pressure reducing valve is calculated as,

$$Q = \frac{C_v\sqrt{\Delta p}}{\sqrt{S_g}} = \frac{2\sqrt{25}}{\sqrt{0.9}} = 10.54\,\text{GPM} \tag{8.38}$$

*Example 8.7* The valve coefficient of a gate valve is given by 3.5 LPM/$\sqrt{}$(kPa). The pressure drop across the valve is measured at 125 kPa. Determine the flow rate if the saturated hydraulic fluid ($S_g$ is 0.9) is used.

**Solution**

The power loss across the pressure reducing valve is calculated as,

$$Q = \frac{C_v\sqrt{\Delta p}}{\sqrt{S_g}} = \frac{3.5\sqrt{125}}{\sqrt{0.9}} = 41.25\,\text{LPM} \tag{8.39}$$

**Practice Problem 8.6**

The valve coefficient of a gate valve is given by 1.5 GPM/$\sqrt{}$(psi). The pressure drop across the valve is measured at 15 psi. Find the flow rate if the saturated hydraulic fluid ($S_g$ is 0.9) is used.

**Practice Problem 8.7**

The valve coefficient of a gate valve is given by 2.5 LPM/$\sqrt{}$(kPa). The pressure drop across the valve is measured at 150 kPa. Determine the flow rate if the saturated hydraulic fluid ($S_g$ is 0.9) is used.

**Exercise Problems**

8.1    The power loss across the pressure relief valve is found to be 25 kW. It returns all the flow from a 0.00011 m³/s pump to the tank. Calculate the pressure setting of a pressure relief valve.

8.2    A pressure relief valve returns all the flow to the tank from a 25 GPM pump. If the power loss across the valve is 23 HP, find the pressure setting of the pressure relief valve.

8.3    The area of the poppet of a pressure relief valve is found 2.14 in². A spring constant is set to 1200 lb/in to hold the poppet at the seat. The adjustment knob is then set to 0.15 in initially from its free length condition. The valve is used to pass full pump flow to the tank at the valve pressure setting, the poppet must be moved 0.06 in from its fully closed position. Find the cracking pressure and the full pump flow pressure.

8.4 The pressure at the primary part of a hydraulic circuit is found to be 1000 psi. Then the pressure is reduced to 600 psi by a pressure reducing valve for the second circuit at a constant flow rate. Find the flow rate if the power loss across the valve is 8 HP.

8.5 The pressure at the primary part of a hydraulic circuit is found to be 1200 Pa. Then the pressure is reduced to a certain value by a pressure reducing valve for the second circuit at a constant flow rate of 25 LPM. Calculate the pressure at the second circuit if the power loss across the valve is 4 kW.

8.6 The valve coefficient of a gate valve is given by 1.4 GPM/$\sqrt{}$(psi). The pressure drop across the valve is measured at 12 psi. Find the flow rate if the saturated hydraulic fluid ($S_g$ is 0.9) is used.

8.7 The valve coefficient of a gate valve is given by 3.3 LPM/$\sqrt{}$(kPa). The pressure drop across the valve is measured at 200 kPa. Find the flow rate if the saturated hydraulic fluid ($S_g$ is 0.9) is used.

# References

1. J.A. Sullivan, *Fluid Power Theory And Applications*, 3rd edn. (Prentice-Hall New Jersey, USA, 1989), pp. 1–528
2. A. Esposito, *Fluid Power with Applications*, 7th edn. (Pearson New International Education, USA, 2010), pp. 1–648
3. F. Don Norvelle, *Fluid Power Technology*, 1st edn. (Delmar, a division of Thomson Learning, USA, 1996), pp. 1–649
4. J.L. Johnson, *Introduction to Fluid Power*, 1st edn. (Delmar, a division of Thomson Learning, USA, 2002), pp. 1–502

# Chapter 9
# Basics of Pneumatics

## 9.1 Introduction

In Greek, the word 'Pneuma' means air. The system that is used gas to transmit power and control energy from one source to another source is known as a pneumatic system. The gas may be compressed air, nitrogen, or any other gases. The air is available in the atmosphere and compressed air is the air from the atmosphere which is reduced in volume by compression that increases its working pressure. Therefore, it is available everywhere without any cost and needs very little storage and delivery costs. Compressed air is used in every application as a safe gas media and it does not affect by temperature or radiation. It does not ignite any fire hazards and does not get hands dirty during application. It does not mess up the equipment and the working space if there is a leaking in the system. A reliable and large volume can be stored using a small container. There are many applications of a pneumatic system such as the manufacturing industry, construction sectors, material handling, airline industry, etc. In this chapter, components of a pneumatic system, comparison, absolute pressure and temperature, gas laws, compressor, and receiver size are discussed.

## 9.2 Components of Pneumatic System

The products or parts that are used compressed air as the medium for operation are called the components of the pneumatic system. A pneumatic system consists of an air compressor, electric motor, filter, receiver, cooler, pressure switch, directional control valve, and cylinder. The filter is used to separate undesirable particles or contaminants from atmospheric air before entering the compressor. The electric motor transforms or converts electrical energy into mechanical energy. Later, this mechanical energy is used to run or drive the air compressor.

The compressor converts the mechanical energy into the potential energy of compressed air. The compressor decreases the volume of the incoming air and

© The Author(s), under exclusive license to Springer Nature Singapore Pte Ltd. 2022    315
Md. A. Salam, *Fundamentals of Pneumatics and Hydraulics*,
https://doi.org/10.1007/978-981-19-0855-2_9

increases the temperature of the air. As the temperature increases, the pressure of the compressed air also increases. Practically, it is difficult to use compressed air at high temperatures. It needs to cool before using in any pneumatic system.

The cooler unit uses water to concentrate and banish excess heat from the compacted or compressed air. Another important component is the receiver or tank. The purpose of the receiver is to provide a smooth flow from the compressor as well as to cool and condense the moisture content. The receiver should be large in volume to hold all compressed air coming out from the compressor. The pressure in the receiver is always kept higher than the system operating pressure to compensate for the pressure loss in the pipe. Also, the larger area of the receiver helps to dissipate the heat of the compressed air. The pressure switch is used to start and stop the electric motor when the system pressure falls and reaches the required pressure level. The operation of a cylinder is controlled by a directional control valve (DCV).

The inputs of a DCV are connected to a pressure line and exhaust or silencer. The outputs (A and B terminals) of a DCV are connected to the cylinder. The components of a pneumatic system are shown in Fig. 9.1.

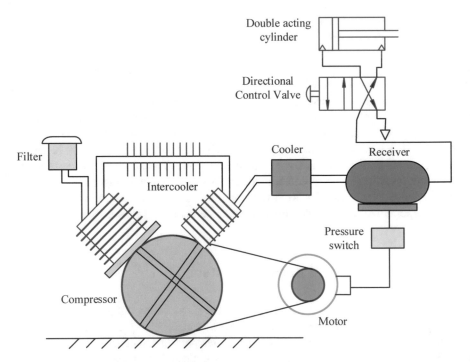

**Fig. 9.1**  A schematic of a pneumatic system

**Table 9.1**  Comparison of pneumatic system

| Pneumatic system | Hydraulic system |
|---|---|
| It uses compressed gas (air) as fluid | It uses pressurized oil as the fluid |
| It is designed for an open system | It is designed for a closed system |
| Leakage does not affect the system | Leakage slows down the system |
| The system is safe and free from fire hazards | The system is not safe as it is very sensitive to fire hazards |
| A special arrangement is required for lubrication | The system is self-lubricated |
| Pressures range from 80 to 100 psi | Pressures range from 1000 to 5000 psi |

## 9.3  Comparison of Pneumatic Systems with Hydraulic Systems

The pneumatic system and hydraulic system are used based on the requirement. In an industry, there are some applications where the pneumatic is used in one part of the circuit and the part of the circuit used a hydraulic system to make the system more efficient and reliable. The comparisons of pneumatic and hydraulic systems are summarized in Table 9.1.

## 9.4  Absolute Pressure and Temperature

The physical force exerted on an object is known as pressure. Mathematically, it is defined as the force applied perpendicular to the object per unit area is known as pressure. The earth is full of air and the air has weight. The air exerts pressure on the earth's surface. The weight exerted by the atmospheric air on a unit area of the earth's surface is known as atmospheric pressure. This atmospheric pressure is measured by a mercury barometer which consists of a glass tube full of mercury. When the tube is inverted over a container, the mercury comes out from the tube and creates the equilibrium vacuum on the top as shown in Fig. 9.2. Therefore, the pressure at point A and point B are equal. The pressure at point B is atmospheric whereas the pressure at point A is the pressure due to mercury column height. The expression of atmospheric pressure is written as,

$$p_{atm} = \rho g h \tag{9.1}$$

For mercury, substituting the values of $\rho = 13,600$ kg/m3, $g = 9.81$ m/s2, $h = 760$ mm at sea level in Eq. (9.1) yields,

**Fig. 9.2** A barometer with a container

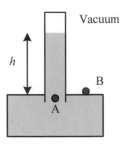

$$p_{atm} = 13600 \times 9.81 \times 0.760 = 101396\,\text{N/m}^2 = 1.013\,\text{atm} = 1.013\,\text{bar} \quad (9.2)$$

The atmospheric pressure decreases exponentially with the increase in altitude. The standard atmospheric pressure in the SI unit is 101.32r kPa and in the Imperial system is 14.7 psi. The pressure above one atmosphere is positive and below one atmosphere is the vacuum. The pressure below the atmospheric pressure is known as vacuum pressure. The pressure measured by a pressure gauge in which atmospheric pressure is taken as a reference is known as gauge pressure and it is represented as psig. In brief, the pressure above the atmospheric pressure is known as gauge pressure. The pressure below the atmospheric pressure is known as vacuum pressure or negative gauge pressure. The relationship between the absolute, atmospheric and gauge pressures is shown in Fig. 9.3. The pressure above absolute zero is known as absolute pressure. Mathematically, the sum of the gauge pressure and atmospheric pressure is known as absolute pressure (abs). The mathematical expression of absolute pressure is,

$$p_{abs} = p_{gauge} + p_{atm} \quad (9.3)$$

**Fig. 9.3** Relationship between pressures

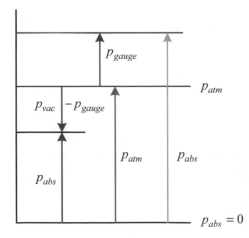

Temperature is a measure of the thermal energy in a body. This can also be expressed as the relative hotness or coldness of a medium and is normally measured in degrees using one of the following scales; Celsius or Centigrade (C), Kelvin (K), Fahrenheit (F) or Rankine (R). In the mid-1700s, Swedish Astronomer Anders Celsius and another by the French Jean Pierre Cristin created the temperature readings of 0° and 100°, giving a 100° scale, for the freezing and boiling points of pure water at atmospheric pressure. In the early 1800s by a German physicist, Daniel Gabriel Fahrenheit proposed the Fahrenheit scale for the temperature. The two points of reference are 32 °F and 212 °F as the freezing and boiling points respectively, of pure water at atmospheric pressure. In the nineteenth century, Scottish engineer William Rankine created the temperature measurement scale shortly after the creation of the Kelvin scale. According to his name, it is known Rankine scale and temperatures can be converted from Fahrenheit to Rankine by adding 459.67. He also noted that the water freezes at 491.67 °R, and boils at 671.67 °R. The temperature in degree Ferenheight is used in the USA. In the US customer system of units, the absolute temperature is measured in degree Rankine and the temperature in degree Ferenheight is converted as,

$$°R =° F + 460 \tag{9.4}$$

In the SI unit, the temperature is expressed in degree Kelvin and the temperature in degree Centigrade is converted as,

$$°K =° C + 273 \tag{9.5}$$

## 9.5 Gas Laws

The pneumatic system is operated by compressed gas. Gases have various properties such as gas pressure, temperature, mass, and the volume which contains the gas. It is important to understand the behaviour of gases. Some experiments were conducted by scientists Boyle, Charles and Gau-Lussac and found that the characteristics of gas follow the law known as ideal gas law.

### 9.5.1 Boyle's Law

Boyle's law, also called Mariotte's law, is a relation to the compression and expansion of a gas at a constant temperature. In 1662, Ireland physicist Robert Boyle formulated a law that the pressure ($p$) of a given quantity of gas varies inversely with its volume ($V$) at a constant temperature (known as the isothermal process). The process which

**Fig. 9.4** A cylinder with a plunger

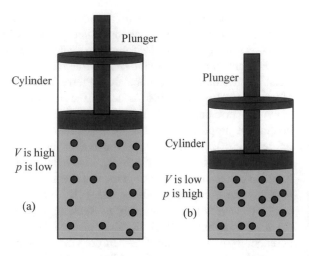

maintains the constant temperature is known as the isothermal process. The relationship was also discovered by the French physicist Edme Mariotte in 1676. Boyle's considered the cylinder and piston with different volumes as shown in Fig. 9.4. In the first cylinder of Fig. 9.4a, the volume is high where the pressure of the gas is low. Similarly, in the second cylinder, these parameters are opposite which means the volume is low and the pressure is high [1]. Therefore, it is seen that the pressure is inversely proportional to the volume of the gas. Mathematically, it can be expressed as,

$$p \propto \frac{1}{V} \tag{9.6}$$

$$pV = k \tag{9.7}$$

For the two gas levels, Eq. (9.7) is revised as,

$$p_1 V_1 = p_2 V_2 \tag{9.8}$$

$$\frac{p_1}{p_2} = \frac{V_2}{V_1} \tag{9.9}$$

where the suffixes represent the initial and final level of the gas.

***Example 9.1*** A cylinder with a volume of 15 in$^3$ contains air at standard atmospheric pressure. The piston is pushed into the cylinder to reduce the volume 9 in$^3$. Find the gauge pressure after the compression.

**Solution**

The absolute pressure at the initial state is calculated as,

$$P_1 = P_{abs} = P_{gauge} + P_{atm} = 0 + 14.7 = 14.7 \, \text{psi} \qquad (9.10)$$

The absolute pressure at the final state is calculated as,

$$P_2 = \frac{P_1 V_1}{V_2} = \frac{14.7 \times 15}{9} = 24.5 \, \text{psia} \qquad (9.11)$$

The gauge pressure is calculated as,

$$P_2 = P_{abs} - P_{atm} = 24.5 - 14.7 = 9.8 \, \text{psig} \qquad (9.12)$$

*Example 9.2* The piston diameter and bore length of a pneumatic cylinder are 1.5 in and 18 in, respectively.

Initially, the cylinder is in a retracted position. In this position, the pressure and volume are found 30 psig and 15 in$^3$. The cylinder needs to retract and it is found that the port is blocked at the cap-end. Calculate the new pressure, assuming the temperature remains constant.

**Solution**

The absolute pressure at the initial state is calculated as,

$$P_1 = P_{abs} = P_{gauge} + P_{atm} = 30 + 14.7 = 44.7 \, \text{psi} \qquad (9.13)$$

The volume at the initial level is,

$$V_1 = 15 \, \text{in}^3 \qquad (9.14)$$

The volume at the final level is calculated as,

$$V_2 = \pi \frac{1.5^2}{4} \times 18 - 15 = 16.81 \, \text{in}^3 \qquad (9.15)$$

The absolute pressure at the final state is calculated as,

$$P_2 = \frac{P_1 V_1}{V_2} = \frac{44.7 \times 15}{16.81} = 39.89 \, \text{psia} \qquad (9.16)$$

The gauge pressure is calculated as,

$$P_2 = P_{abs} - P_{atm} = 39.89 - 14.7 = 25.19 \, \text{psig} \qquad (9.17)$$

**Practice Problem 9.1**

Air is stored in a cylinder with a volume of 3 m$^3$ at a pressure of 125 $^{kPa}$. If the air is compressed to a volume of 1.2 m$^3$, find the gauge pressure in kPa after compression.

**Practice Problem 9.2**

The piston diameter and bore length of a pneumatic cylinder are 2 in and 15 in, respectively. A pressure of 25 psi is required to push the workpiece by the cylinder. Find the pressure at this position if the rod diameter is 1 in. Assuming the temperature remains constant.

## 9.5.2  Charles's Law

A French physicist JACQVES Charles (1746–1823) deals with the changing of gas from one temperature and volume condition to another while holding the pressure constant. This law is known as Charles law, also known as a perfect gas law. The law states that at constant pressure, the volume $V$ of a gas is directly proportional to its absolute (Kelvin) temperature $T$. A burner is used to heat the cylinder in Fig. 9.5a which in turn increases the temperature and volume as can be seen in Fig. 9.5b. Mathematically, this law can be expressed as,

$$V \propto T \tag{9.18}$$

$$V = k_1 T \tag{9.19}$$

$$\frac{V}{T} = k_1 \tag{9.20}$$

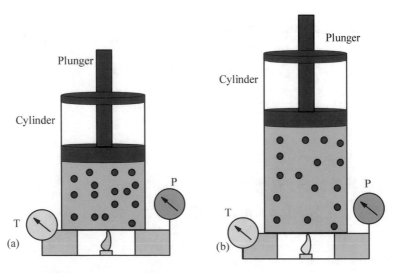

**Fig. 9.5**  A cylinder with measuring instruments

For initial and the final states of the gas, Eq. (9.20) is revised as,

$$\frac{V_1}{T_1} = \frac{V_2}{T_2} \tag{9.21}$$

**Example 9.3** The initial temperature of the air in a container of 15 in³ volume is found at 55 °F. If the air is heated to 240 °F, determine its final volume, assuming that the pressure remains constant.

**Solution**

The initial temperature is calculated as,

$$T_1 = 55 + 460 = 515°R \tag{9.22}$$

The final temperature is calculated as,

$$T_2 = 240 + 460 = 700°R \tag{9.23}$$

The final volume is calculated as,

$$V_2 = T_2 \frac{V_1}{T_1} = 700 \times \frac{15}{515} = 20.39 \, in^3 \tag{9.24}$$

**Practice Problem 9.3**

The initial temperature of the air in a 0.5 m³ volume container is found at 35 °C. If the air is heated to 140 °C, determine its final volume, assuming that the pressure remains constant.

### 9.5.3 Gay-Lussac's Law

In 1802, French chemist and physicist Joseph Louis Gay-Lussac discovered that if the volume of gas keeps constant in a closed container and if apply heat, the pressure of the gas will increase. As a result, the gases will have more kinetic energy, causing them to hit the walls of the container with more force which in turn produces greater pressure as shown in Fig. 9.6. Gay-Lussac's law deals with the changing of gas from one pressure and temperature condition to another while holding the volume constant. When the volume of a gas is held constant, the pressure exerted by the gas is directly proportional to the absolute temperature. Mathematically, it is expressed as [2],

$$p \propto T \tag{9.25}$$

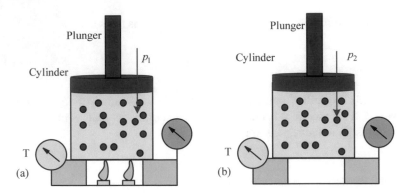

**Fig. 9.6** Two cylinders with different pressures

$$\frac{p}{T} = k_2 \tag{9.26}$$

For initial and final levels of the gas, Eq. (9.26) is revised as,

$$\frac{p_1}{T_1} = \frac{p_2}{T_2} \tag{9.27}$$

***Example 9.4*** A 0.56 m³ volume receiver filled with compressed air at 16 psi (gauge) at an initial temperature of 50 °C. If the temperature cools to 30 °C, find the final gauge pressure.

**Solution**

The initial pressure is calculated as,

$$p_1 = p_{atm} + p_{gauge} = 14.7 + 16 = 30.7 \, \text{psi(abs)} \tag{9.28}$$

The initial and final temperatures are calculated as,

$$T_1 = 50 + 273 = 323°\text{K} \tag{9.29}$$

$$T_2 = 30 + 273 = 303°\text{K} \tag{9.30}$$

The final pressure is calculated as,

$$p_2 = \frac{p_1}{T_1} \times T_2 = 30.7 \times \frac{303}{323} = 28.80 \, \text{psi(abs)} \tag{9.31}$$

The final pressure in gauge is calculated as,

$$p_2 = p_{abs} - p_{atm} = 28.80 - 14.7 = 14.1\,\text{psi(gauge)} \qquad (9.32)$$

**Practice Problem 9.4**

Air is stored in a fixed volume container at atmospheric pressure (0 psig) and 0 °C. If the temperature rises to 130 °C, find the final gauge pressure.

## 9.5.4 General Gas Law

It is seen that any given mass of gas undergoes changes in pressure, volume and temperature. The general gas law is obtained by combining Boyle's, Charle's and Gay-Lussac's laws as,

$$\frac{p_1 V_1}{T_1} = \frac{p_2 V_2}{T_2} \qquad (9.33)$$

The pressure, volume and temperature are not held constant during a process from the initial level to the final level. The value of any parameters can be calculated if the values of other parameters are given in a general gas law.

**Example 9.5** A gas of 65 psi (gauge) is stored in a 950 cm$^3$ volume cylinder at a temperature of 50 $^\circ$C. The piston of the cylinder compressed the volume to 620 cm$^3$ while the gas is heated to 90 °C. Calculate the final gauge pressure.

**Solution**

The initial pressure is calculated as,

$$p_1 = p_{atm} + p_{gauge} = 14.7 + 65 = 79.7\,\text{psi(abs)} \qquad (9.34)$$

The initial and final temperatures are calculated as,

$$T_1 = 50 + 273 = 323^\circ\text{K} \qquad (9.35)$$

$$T_2 = 90 + 273 = 363^\circ\text{K} \qquad (9.36)$$

The final pressure is calculated as,

$$p_2 = \frac{T_2}{V_2} \times \frac{p_1 V_1}{T_1} = \frac{363}{620} \times \frac{79.7 \times 950}{323} = 137.24\,\text{psi(abs)} \qquad (9.37)$$

The final pressure in gauge is calculated as,

$$p_2 = p_{abs} - p_{atm} = 137.24 - 14.7 = 122.54\text{psi(gauge)} \qquad (9.38)$$

**Practice Problem 9.5**

A gas of 25 psi (gauge) is stored in a 1250 cm³ volume cylinder at a temperature of 85 °C. The piston of the cylinder compressed the volume to 1000 cm³ while the gas is heated to 130 °C. Calculate the final gauge pressure.

## 9.6 Compressor Isothermal and Adiabatic Process

The proper size of the compressor is very important to provide maximum airflow with maximum pressure to the fluid power system for operating efficiently. The data for minimum, average and maximum airflow, minimum and maximum airflow pressures and the condition are required for designing a proper size of the compressor. In general, work done per cycle by the compressor is required to calculate the power rating of the compressor. The compressor activity mainly depends on the adiabatic process. Here, the isothermal and adiabatic processes are discussed.

### 9.6.1 Isothermal Process

The process in which the temperature is not changed but pressure and volume may change is known as isothermal. The first law of thermodynamics is defined as the change in heat ($dQ$) applied in a system is equal to the sum of the change of energy ($dU$) and change in work ($dW$) done by the system. Mathematically, it is expressed as,

$$dQ = dU + dW \tag{9.39}$$

In an isothermal process, the following relation is written as,

$$dT = 0 \tag{9.40}$$

However, the energy is a function of temperature and it is expressed as,

$$U = f(T) = 0 \tag{9.41}$$

Consider a cylinder with a diathermal wall as shown in Fig. 9.7. This cylinder with a diathermal wall will absorb from the surroundings and reject heat to the surroundings. Therefore, the differential form of energy will be equal to zero. Equation (9.39) becomes,

$$dQ = dW \tag{9.42}$$

**Fig. 9.7** A cylinder with a diathermal wall

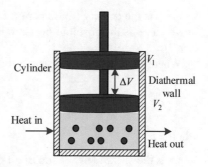

From Eq. (9.42), it is seen that all heat energy applied to a system (isothermal) will be converted to work done. An isothermal process diagram is shown in Fig. 9.8 where the initial volume is $V_1$ and the final volume is $V_2$. The final volume is decreased when the piston of the cylinder pushes down. Therefore, the work done is expressed as,

$$dW = p\Delta V \tag{9.43}$$

The total work done is calculated by integrating Eq. (9.43) with a limit from initial volume to final volume as,

$$W = \int dW = \int_{V_1}^{V_2} p\, dV \tag{9.44}$$

For one mole, the ideal gas law is expressed as [3],

$$\frac{p}{\rho} = pV = RT \tag{9.45}$$

**Fig. 9.8** An isothermal $pV$ diagram

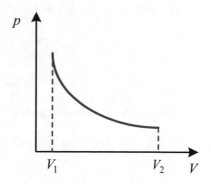

where $\rho$ is the density (mass per unit volume) of gas, $T$ is the absolute temperature in Kelvin, $p$ is the absolute pressure in psi or kPa, $V$ is the volume of the gas in ft$^3$ or m$^3$.

Where $R$ is the universal gas or molar gas constant and it is expressed as,

$$R = C_p - C_v \tag{9.46}$$

where,

$C_p$ is the specific heat at constant pressure,
$C_v$ is the specific heat at constant volume.

Substituting Eq. (9.45) in Eq. (9.44) yields,

$$W = \int dW = \int_{V_1}^{V_2} RT \frac{dV}{V} \tag{9.47}$$

$$W = RT \ln \frac{V_2}{V_1} \tag{9.48}$$

From Eq. (9.48), it is seen that the work done on an isothermal process can be calculated if the volumes and temperature are given.

## 9.6.2  Adiabatic Process

To calculate the work done in a adiabatic process, differentiating Eq. (9.45) yields,

$$pdV + Vdp = RdT \tag{9.49}$$

Substituting Eq. (9.46) into Eq. (9.49) yields,

$$pdV + Vdp = (C_p - C_v)dT \tag{9.50}$$

$$pdV + C_v dT = C_p dT - Vdp \tag{9.51}$$

However, the following equation is written as,

$$C_v = \frac{dU}{dT} \tag{9.52}$$

Substituting Eqs. (9.43) and (9.52) into Eq. (9.39) yields,

**Fig. 9.9** An adiabatic process

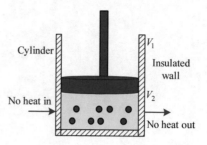

Cylinder

$V_1$

Insulated wall

No heat in

$V_2$

No heat out

$$dQ = C_v dT + pdV \qquad (9.53)$$

The adiabatic process is the process that no heat is coming in and going out that means $dQ$ is equal to zero as shown in Fig. 9.9. Therefore, Eqs. (9.53) and (9.51) can be written as,

$$C_v dT + pdV = 0 \qquad (9.54)$$

$$C_p dT - Vdp = 0 \qquad (9.55)$$

Equations (9.54) and (9.55) can be revised as,

$$C_v dT = -pdV \qquad (9.56)$$

$$C_p dT = Vdp \qquad (9.57)$$

Dividing Eq. (9.56) by (9.57) yields,

$$-\frac{C_v}{C_p} = \frac{dV}{V} \times \frac{p}{dp} \qquad (9.58)$$

$$-\frac{dp}{p} = \frac{C_p}{C_v} \times \frac{dV}{V} \qquad (9.59)$$

The adiabatic ratio is defined as,

$$\gamma = \frac{C_p}{C_v} \qquad (9.60)$$

Substituting Eq. (9.60) into Eq. (9.59) yields,

$$-\frac{dp}{p} = \gamma \frac{dV}{V} \qquad (9.61)$$

Integrating Eq. (9.59) yields,

$$\ln k - \ln p = \gamma \ln V \tag{9.62}$$

$$\ln k = \ln p + \ln V^{\gamma} \tag{9.63}$$

$$\ln k = \ln p V^{\gamma} \tag{9.64}$$

$$p V^{\gamma} = k \tag{9.65}$$

Based on Eq. (9.65), for initial and final levels, the adiabatic equation is expressed as,

$$p_1 V_1^{\gamma} = p_2 V_2^{\gamma} = k \tag{9.66}$$

From Eq. (9.65), it is seen that the pressure will increase with a lower volume as a power of one. The adiabatic process has a larger slope either in expansion or compression as shown in Fig. 9.10. Substituting Eq. (9.45) into Eq. (9.65) yields,

$$\frac{RT}{V} V^{\gamma} = k \tag{9.67}$$

$$T(V)^{\gamma-1} = \frac{k}{R} = k_1 \tag{9.68}$$

For two levels of a system, Eq. (9.68) can be expanded as,

$$T_1(V_1)^{\gamma-1} = T_2(V_2)^{\gamma-1} = k_1 \tag{9.69}$$

$$\frac{T_1}{T_2} = \left(\frac{V_2}{V_1}\right)^{\gamma-1} \tag{9.70}$$

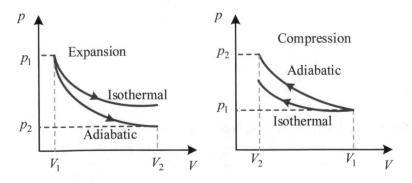

**Fig. 9.10** An adiabatic process $pV$ diagram

$$\frac{V_2}{V_1} = \left(\frac{T_1}{T_2}\right)^{\frac{1}{\gamma-1}} \tag{9.71}$$

Equation (9.66) can be rearranged as,

$$\frac{p_1}{p_2} = \left(\frac{V_2}{V_1}\right)^{\gamma} \tag{9.72}$$

Substituting Eq. (9.71) into Eq. (9.72) yields,

$$\frac{p_1}{p_2} = \left(\frac{T_1}{T_2}\right)^{\frac{\gamma}{\gamma-1}} \tag{9.73}$$

Substituting $\gamma = 1.4$ for air and Eq. (9.73) yields,

$$\frac{p_1}{p_2} = \left(\frac{T_1}{T_2}\right)^{3.5} \tag{9.74}$$

The work done for a thermodynamic process is calculated as,

$$W_{12} = \int_{V_1}^{V_2} p \, dV \tag{9.75}$$

Substituting Eq. (9.65) into Eq. (9.75) yields,

$$W_{12} = \int_{V_1}^{V_2} \frac{k}{V^{\gamma}} \, dV \tag{9.76}$$

$$W_{12} = (k)\left[\frac{V^{-\gamma+1}}{1-\gamma}\right]_{V_1}^{V_2} \tag{9.77}$$

$$W_{12} = \frac{k}{1-\gamma}\left[V_2^{-\gamma+1} - V_1^{-\gamma+1}\right] \tag{9.78}$$

Substituting the value of $k$ from Eq. (9.66) into Eq. (9.78) yields,

$$W_{12} = \frac{1}{1-\gamma}\left[p_2 V_2^{\gamma}(V_2^{-\gamma+1}) - p_1 V_1^{\gamma}(V_1^{-\gamma+1})\right] \tag{9.79}$$

$$W_{12} = \frac{1}{1-\gamma}[p_2 V_2 - p_1 V_1] \tag{9.80}$$

According to Eq. (9.39), the change in energy for an adiabatic process ($dQ = 0$) is expressed as,

$$0 = dU + dW \tag{9.81}$$

$$dU = -dW \tag{9.82}$$

From Eq. (9.82), it is seen that the change in energy will be positive for a negative value of the change in work done. Therefore, the work done under compression is expressed as,

$$W_{12} = -\frac{1}{1-\gamma}[p_2 V_2 - p_1 V_1] \tag{9.83}$$

$$W_{12} = \frac{1}{\gamma - 1}[p_2 V_2 - p_1 V_1] \tag{9.84}$$

The cylinder with an induction or suction port and delivery port is shown in Fig. 9.11. The induction phase of the cylinder is represented by 4–1, the adiabatic phase is represented by 1–2 and the delivery phase is represented by 2–3 as shown in Fig. 9.11.

The area under the p–V diagram is the net work done on the air per cycle. Therefore, the work done by the cylinder is equal to the area under 1–2–3–4–1. Mathematically, it is expressed as,

**Fig. 9.11** A complete cycle of an air induction and delivery process

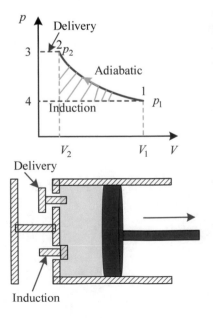

$$W = \text{area under } 1-2 + \text{area under } 2-3 - \text{area under } 4-1 \qquad (9.85)$$

Substituting Eq. (9.84) as the area under 1–2, $p_2 V_2$ as the area under 2–3, $p_1 V_1$ as the area under 4–1 in Eq. (9.85) yields,

$$W = \frac{1}{\gamma - 1}[p_2 V_2 - p_1 V_1] + p_2 V_2 - p_1 V_1 \qquad (9.86)$$

$$W = p_2 V_2 \left( \frac{1}{\gamma - 1} + 1 \right) - p_1 V_1 \left( \frac{1}{\gamma - 1} + 1 \right) \qquad (9.87)$$

$$W = \frac{\gamma}{\gamma - 1} p_2 V_2 - \frac{\gamma}{\gamma - 1} p_1 V_1 \qquad (9.88)$$

$$W = \frac{\gamma}{\gamma - 1} p_1 V_1 \left( \frac{p_2 V_2}{p_1 V_1} - 1 \right) \qquad (9.89)$$

Equation (9.72) can be rewritten as,

$$\frac{V_2}{V_1} = \left( \frac{p_1}{p_2} \right)^{\frac{1}{\gamma}} \qquad (9.90)$$

Substituting Eq. (9.90) into Eq. (9.89) yields,

$$W = \frac{\gamma}{\gamma - 1} p_1 V_1 \left[ \frac{p_2}{p_1} \left( \frac{p_1}{p_2} \right)^{\frac{1}{\gamma}} - 1 \right] \qquad (9.91)$$

$$W = \frac{\gamma}{\gamma - 1} p_1 V_1 \left[ \frac{\left( \frac{p_2}{p_1} \right)}{\left( \frac{p_2}{p_1} \right)^{-\frac{1}{\gamma}}} - 1 \right] \qquad (9.92)$$

$$W = \frac{\gamma}{\gamma - 1} p_1 V_1 \left[ \left( \frac{p_2}{p_1} \right)^{1 - \frac{1}{\gamma}} - 1 \right] \qquad (9.93)$$

$$W = \frac{\gamma}{\gamma - 1} p_1 V_1 \left[ \left( \frac{p_2}{p_1} \right)^{\frac{\gamma - 1}{\gamma}} - 1 \right] \qquad (9.94)$$

The power is defined as the rate of receiving or delivering work and it can be expressed as,

$$P = \frac{dW}{dt} \qquad (9.95)$$

In general, the expression of power is expressed as,

$$P = \frac{W}{t} \tag{9.96}$$

Again, the extension or retraction time of a cylinder is defined as the ratio of the volume flow rate to the flow rate of the fluid. Mathematically, it is expressed as,

$$t = \frac{V}{Q} \tag{9.97}$$

Substituting Eq. (9.94) and (9.97) into Eq. (9.96) yields the theoretical power,

$$P_T = \frac{\frac{\gamma}{\gamma-1} p_1 V_1 \left[ \left( \frac{p_2}{p_1} \right)^{\frac{\gamma-1}{\gamma}} - 1 \right]}{\frac{V}{Q}} \tag{9.98}$$

$$P_T = \frac{\gamma}{\gamma-1} p_1 Q \left[ \left( \frac{p_2}{p_1} \right)^{\frac{\gamma-1}{\gamma}} - 1 \right] \tag{9.99}$$

For air, the value of $\gamma = 1.4$, then Eq. (9.99) is modified as,

$$P_T = 3.5 p_1 Q \left[ \left( \frac{p_2}{p_1} \right)^{0.286} - 1 \right] \tag{9.100}$$

where,

$p_1$ is the inlet pressure in psi, kPa abs,
$p_2$ is the outlet pressure in psi, kPa abs,
$P_T$ is the theoretical power,
$Q$ is the flow rate in scfm, standard ft³/min, m³/min.

Use 1 HP = 33,000 ft-lb/min to convert the power of Eq. (9.100) in HP as,

$$P_T(\text{HP}) = 3.5 \frac{(p_1 \times 12 \times 12 \, \text{lb/ft}^2) Q(\text{ft}^3/\text{min})}{33000} \left[ \left( \frac{p_2}{p_1} \right)^{0.286} - 1 \right] \tag{9.101}$$

$$P_T(\text{HP}) = \frac{p_1 Q}{65.5} \left[ \left( \frac{p_2}{p_1} \right)^{0.286} - 1 \right] \tag{9.102}$$

The expression of apparent theoretical power in SI unit can be derived as,

$$1 \, \text{W} = 1 \, \text{N} - \text{m/s} = 60 \, \text{N} - \text{m/min} \tag{9.103}$$

$$1 \, \text{N} - \text{m/min} = \frac{1}{60} \, \text{W} = \frac{1}{60 \times 1000} \text{kW} \tag{9.104}$$

In SI unit, Eq. (9.100) can be written as,

$$P_T = 3.5\,(p_1 \times 10^3\,\text{N/m}^2)\,Q\,(\text{m}^3/\text{min})\left[\left(\frac{p_2}{p_1}\right)^{0.286} - 1\right] \tag{9.105}$$

Substituting Eq. (9.104) into Eq. (9.105) yields the theoretical power in kW,

$$P_T\,(\text{kW}) = 3.5\,\frac{(p_1 \times 10^3\,\text{N/m}^2)\,Q\,(\text{m}^3/\text{min})}{60 \times 1000}\left[\left(\frac{p_2}{p_1}\right)^{0.286} - 1\right] \tag{9.106}$$

$$P_T\,(\text{kW}) = \frac{p_1 Q}{17.14}\left[\left(\frac{p_2}{p_1}\right)^{0.286} - 1\right] \tag{9.107}$$

The overall compressor efficiency is defined as the ratio of theoretical power ($P_T$) to the apparent power ($P_A$) and mathematically, it is expressed as,

$$\eta_0 = \frac{P_T}{P_A} \tag{9.108}$$

Substituting Eqs. (9.102) and (9.107) into Eq. (9.108) yields the actual power,

$$P_A\,(\text{HP}) = \frac{p_1 Q}{\eta_0 \times 65.5}\left[\left(\frac{p_2}{p_1}\right)^{0.286} - 1\right] \tag{9.109}$$

$$P_A\,(\text{kW}) = \frac{p_1 Q}{\eta_0 \times 17.14}\left[\left(\frac{p_2}{p_1}\right)^{0.286} - 1\right] \tag{9.110}$$

*Example 9.6* In an adiabatic process, the air is compressed from 6 ft$^3$ to 1.5 ft$^3$ at 60 °F and atmospheric pressure. Calculate the final pressure and temperature of the gas.

**Solution**

The temperature in °R is calculated as,

$$T_1 = 60 + 460 = 520°R \tag{9.111}$$

The final pressure is calculated as,

$$p_2 = \frac{p_1 V_1^{1.4}}{V_2^{1.4}} = \frac{14.7 \times 6^{1.4}}{1.5^{1.4}} = 102.38\,\text{psia} \tag{9.112}$$

The final pressure in psig is calculated as,

$$p_2 = p_{abs} - p_{atm} = 102.38 - 14.7 = 87.68 \, \text{psig} \tag{9.113}$$

The final temperature is calculated as,

$$T_2 = \sqrt[3.5]{\frac{p_2}{p_1}} \times T_1^{3.5} = \sqrt[3.5]{\frac{102.38}{14.7}} \times 520^{3.5} = 905.38°\text{R} \tag{9.114}$$

$$T_2 = 905.38 - 460 = 445.33°\text{F} \tag{9.115}$$

**Practice Problem 9.6**
In an adiabatic process, the air is compressed from 5 m³ to 0.86 m³ at 30 °C and atmospheric pressure. Determine the final pressure and temperature of the gas.

***Example 9.7*** A compressor delivers 80 scfm of air at 90 psig. The overall efficiency of a compressor is 80%. Determine the theoretical power and apparent power.

**Solution**

The final pressure is calculated as,

$$p_2 = p_{atm} + p_{gauge} = 14.7 + 90 = 104.7 \, \text{psia} \tag{9.116}$$

The theoretical power is calculated as,

$$P_T(\text{HP}) = \frac{p_1 Q}{65.5}\left[\left(\frac{p_2}{p_1}\right)^{0.286} - 1\right] = \frac{14.7 \times 80}{65.5}\left[\left(\frac{104.7}{14.7}\right)^{0.286} - 1\right] = 13.52 \, \text{HP} \tag{9.117}$$

The apparent power is calculated as,

$$P_A(\text{HP}) = \frac{p_1 Q}{\eta_0 \times 65.5}\left[\left(\frac{p_2}{p_1}\right)^{0.286} - 1\right] = \frac{13.52}{0.8} = 16.91 \, \text{HP} \tag{9.118}$$

**Practice Problem 9.7**
The compressor delivers 8 standard m³/min at 750 kPa. The overall efficiency of a compressor is 75%. Determine the theoretical power and apparent power.

## 9.7  Receiver

The main function of a receiver is used to store air with high pressure. The receiver is often known as a tank. Besides storing the high-pressure air, the receiver determines

**Fig. 9.12** A receiver with a cylinder

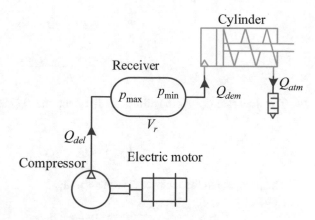

the exact amount of air needed to run the fluid power system efficiently. Consider a compressor delivers a flow $Q_{del}$ to the receiver and the single-acting cylinder demands the flow $Q_{dem}$ to operate as shown in Fig. 9.12.

The maximum and minimum pressures of the air at the receiver are $p_{max}$ and $p_{min}$, respectively. The compressor is on when the pressure reduces to the minimum value and fills the receiver with high-pressure air. The is flow rate usually goes to the atmosphere after operating the cylinder. Consider Boyle's gas law to derive the size of the receiver.

Substituting equation $p1 = p_r, V_1 = V_r, p_2 = p_{atm}, V_2 = V_{atm}$ in Eq. (9.8) as,

$$p_r V_r = p_{atm} V_{atm} \qquad (9.119)$$

where,

$p_r$ is the pressure of the receiver,
$V_r$ is the volume of the receiver,
$p_{atm}$ is the pressure of the atmosphere,
$V_{atm}$ is the volume of the atmosphere.

Dividing Eq. (9.119) by $t$ yields,

$$p_r \frac{V_r}{t} = p_{atm} \frac{V_{atm}}{t} \qquad (9.120)$$

The flow rate ($Q$) is defined as the volume flow rate ($V$) per unit cycle time ($t$) as,

$$Q = \frac{V}{t} \qquad (9.121)$$

Substituting Eq. (9.121) into Eq. (9.120) yields,

$$p_r \frac{V_r}{t} = p_{atm} \, Q_{atm} \tag{9.122}$$

$$V_r = \frac{p_{atm} \times t \times Q_{atm}}{p_r} \tag{9.123}$$

The flow rate $Q_{atm}$ is the difference between the $Q_{del}$ and $Q_{dem}$ and it is expressed as,

$$Q_{atm} = Q_{dem} - Q_{del} \tag{9.124}$$

The pressure in the receiver is expressed as,

$$p_r = p_{max} - p_{min} \tag{9.125}$$

Substituting Eqs. (9.124) and (9.125) into Eq. (9.123) yields,

$$V_r = \frac{p_{atm} \times t \times (Q_{dem} - Q_{del})}{p_{max} - p_{min}} \tag{9.126}$$

where,

$p_{max}$ is the maximum pressure of the receiver in psig, kPa,
$V_r$ is the volume of the receiver, ft$^3$, m$^3$,
$p_{min}$ is the minimum pressure of the receiver in psig, kPa,
$Q_{dem}$ is the demand flow rate of the fluid power system in scfm, m$^3$/min,
$Q_{del}$ is the delivery flow rate by the compressor to the receiver in scfm, m$^3$/min
$p_{atm}$ is the atmospheric pressure in 14.7 psig, 101 kPa.

Equation (9.126) can again be revised by substituting the atmospheric pressure as,

$$V_r = \frac{14.7 \times t \times (Q_{dem} - Q_{del})}{p_{max} - p_{min}} \tag{9.127}$$

$$V_r = \frac{101 \times t \times (Q_{dem} - Q_{del})}{p_{max} - p_{min}} \tag{9.128}$$

**Example 9.8** A pneumatic system demands 40 scfm and the compressor delivers 25 scfm. The receiver runs for 6 min and the pressure reduces from 90 to 80 psi. Determine the receiver size.

**Solution**

The receiver size is calculated as,

$$V_r = \frac{14.7 \times t \times (Q_{dem} - Q_{del})}{p_{max} - p_{min}} = \frac{14.7 \times 6 \times (40 - 25)}{90 - 70} = 66.15 \, \text{ft}^3 \tag{9.129}$$

***Example 9.9*** A pneumatic system demands 50 m³/min and the compressor delivers 30 m³/min. The receiver runs for 10 min and the pressure reduces from 120 to 70 kPa. Calculate the receiver size.

**Solution**

The receiver size is calculated as,

$$V_r = \frac{101 \times t \times (Q_{dem} - Q_{del})}{p_{max} - p_{min}} = \frac{101 \times 10 \times (50 - 30)}{120 - 70} = 404\,\text{m}^3 \quad (9.130)$$

**Practice Problem 9.8**
An air receiver with a size of 5 scfm is charged to 125 psi. The compressor delivers 2 scfm at 85 psi before going to shut down. Calculate the time.

**Practice Problem 9.9**
An air receiver delivers the mean flow rate of 3 m³/min at 550 kPa for 4 min and then the compressor shuts down. Determine the size of the receiver if it is charged at 950 kPa.

# 9.8   Vacuum and Suction Force

There are some applications in a pneumatic system where vacuum pressure is used to get more reliable results. Some of these applications are semiconductor fabrications, plasma sterilizer, vacuum packaging, moulding, clamping, transporting, filling and sealing. The pressure below the atmospheric pressure is known as vacuum pressure. It is measured as pounds per square inch vacuum (PSIV) or 0–30″ Hg, mm of Hg (millimetre of mercury), torr. For a very high vacuum, the pressure is measured in torr. In the seventeenth century, Italian scientist, Evangelista Torricelli, invented the mercury barometer. Based on these, one-atmosphere pressure is expressed as,

$$1\,\text{atm} = 760\,\text{mmHg} = 760\,\text{torr} \quad (9.131)$$

Different ranges of vacuums along with the corresponding pressure ranges are reported in Table 9.2.

Suction is the process of reducing air pressure by removing air from an enclosed surface. The suction force is the magnitude of the force that holds or lifts a specific amount of load.

Let a flat metal needs to be lifted by a suction cup as shown in Fig. 9.13. Let consider the suction force is $F_s$ to lift the metal piece. The atmospheric pressure acts on all sides of the metal piece.

Therefore, the vacuum pressure is lower than the atmospheric pressure. The suction force ($F_s$) is the difference between atmospheric pressure and vacuum

**Table 9.2** Different levels of vacuum

| Vacuum | Pressure |
|---|---|
| Low | 1 atm–1 Pa |
| Medium | 1 Pa–0.1 Pa |
| High | 0.1 Pa–$10^{-5}$ Pa |
| Ultra high | $10^{-5}$ Pa–$10^{-10}$ Pa |
| Extra high | $<10^{-10}$ Pa |

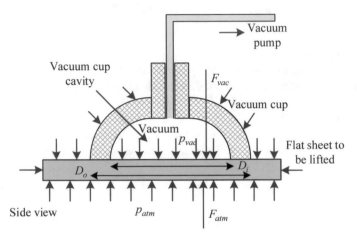

**Fig. 9.13** A side view of the succession process

pressure. Mathematically, it is expressed as,

$$F_s = F_{atm} - F_{vac} \tag{9.132}$$

According to fundamental definition of pressure ($p = F/A$), Eq. (9.132) is revised as,

$$F_s = p_{atm} A_o - p_{vac} A_i \tag{9.133}$$

$$F_s = p_{atm} \frac{\pi D_o^2}{4} - p_{vac} \frac{\pi D_i^2}{4} \tag{9.134}$$

where,

$F_s$ is the upward suction force or lifting force exerted on the metal piece in lb, N,
$p_{atm}$ is the atmospheric pressure in Psia, Pa abs,
$p_{suc}$ or $p_{vac}$ is the suction pressure in Psia, Pa abs,
$A_o$ is the outer circle area of the suction cup in in$^2$, m$^2$,
$A_i$ is the inner circle area of the suction cup in in$^2$, m$^2$,

$D_o$ is the outer circle diameter of the suction cup in in², m²,
$D_i$ is the inner circle diameter of the suction cup in in², m²,

The time to achieve vacuum pressure needs to be calculated. When the atmospheric pressure acts on the suction cups, then the air removes from the inside of the suction cups. Therefore, the vacuum is created underneath the suction cups which in turn changes the pressure. The work done is calculated as,

$$dW = V \, dp \tag{9.135}$$

Substituting Eq. (9.45) into Eq. (9.135) yields,

$$dW = \frac{RT}{p} \, dp \tag{9.136}$$

Integrating Eq. (9.136) from $p_{vac}$ to $p_{atm}$ yields,

$$\int dW = RT \int_{p_{vac}}^{p_{atm}} \frac{dp}{p} \tag{9.137}$$

$$W = RT[\ln p]_{p_{vac}}^{p_{atm}} \tag{9.138}$$

$$W = RT \ln\left(\frac{p_{atm}}{p_{vac}}\right) \tag{9.139}$$

In the previous chapters, the expression of power is written as,

$$P = p \, Q \tag{9.140}$$

Substituting Eq. (9.139) into Eq. (9.96) yields,

$$P = \frac{RT \ln\left(\frac{p_{atm}}{p_{vac}}\right)}{t} \tag{9.141}$$

Substituting Eq. (9.140) into Eq. (9.141) yields,

$$p \, Q = \frac{RT \ln\left(\frac{p_{atm}}{p_{vac}}\right)}{t} \tag{9.142}$$

$$t = \frac{RT}{p} \frac{\ln\left(\frac{p_{atm}}{p_{vac}}\right)}{Q} \tag{9.143}$$

Again, substituting Eq. (9.45) into Eq. (9.143) yields,

$$t = \frac{V}{Q} \ln\left(\frac{p_{atm}}{p_{vac}}\right) \tag{9.144}$$

where,

$t$ is the required time to achieve the pump vacuum pressure in min,
$V$ is the total volume of the suction cups and connecting pipe in ft$^3$, m$^3$,
$Q$ is the flow rate generated by the vacuum pump in ft$^3$/min, m$^3$/min,
$p_{atm}$ is the atmospheric pressure in psia, Pa abs,
$p_{vac}$ is the vacuum pressure in psia, Pa abs.

**Example 9.10**  The inside and outside diameters of suction cups are 4 in and 6 in, respectively. Calculate the lifting force for—9 psig $=$ 9 psi suction $=$ 9 psi vacuum.

**Solution**

The absolute suction pressure is calculated as,

$$p_{suc}(abs) = p_{gauge} + p_{atm} = -9 + 14.7 = 5.7 \, \text{psia} \tag{9.145}$$

The value of the lifting force is calculated as,

$$F_s = p_{atm} \frac{\pi D_o^2}{4} - p_{vac} \frac{\pi D_i^2}{4} = 14.7 \times \frac{\pi \, 6^2}{4} - 5.7 \times \frac{\pi \, 4^2}{4} = 344 \, \text{lb} \tag{9.146}$$

**Example 9.11**  A pneumatic system lifts a workpiece using a total volume of 8 ft$^3$ inside the suction cup and associated pipeline up to the vacuum pump. The atmospheric and the suction pressures are 14.7 psia and 9 psia. Calculate the required time to obtain the required level of vacuum pressure when the pump produces a flow rate of 5 scfm.

**Solution**

The time is calculated as,

$$t = \frac{V}{Q} \ln\left(\frac{p_{atm}}{p_{vac}}\right) = \frac{8}{5} \ln\left(\frac{14.7}{9}\right) = 0.78 \text{min} \tag{9.147}$$

**Practice Problem 9.10**
The inside and outside diameters of suction cups are found to 3.5 cm and 5.5 cm, respectively. Calculate the lifting force for—10 kPa $=$ 10 kPa suction $=$ 10 kPa vacuum.

**Practice Problem 9.11**
A pneumatic system lifts a workpiece using a total volume of 6.5 m$^3$ inside the suction cup and associated pipeline up to the vacuum pump. The atmospheric and the suction pressures are 101 kPa abs and 40 kPa abs. Determine the required time

to obtain the required level of vacuum pressure when the pump produces a flow rate of 4 m$^3$/min.

## Exercise Problems

9.1    A cylinder with a volume of 9 ft$^3$ contains air at standard atmospheric pressure. The piston is pushed into the cylinder to reduce the volume by 2 ft$^3$. Calculate the gauge pressure after the compression.

9.2    Air is stored in a cylinder with a volume of 4.5 m$^3$ at a pressure of 112 $^{kPa}$ gauge. If the air is compressed to a volume of 1.2 m$^3$, determine the gauge pressure in kPa after compression.

9.3    Air is stored in a cylinder with a volume of 1.2 ft$^3$ at a pressure of 80 psig. Then the air is expanded to 4.5 ft$^3$ at a constant temperature, find the gauge pressure in psig after expansion.

9.4    Air is stored in a cylinder with a volume of 0.5 m$^3$ at a pressure of 250 $^{kPa}$ gauge. The air is expanded to a volume of 4.2 m$^3$ at a constant temperature, calculate the gauge pressure in kPa after expansion.

9.5    The initial temperature of the air in a 1.5 m$^3$ volume container is found at 45 °C. Then the air is heated to 125 °C, determine its final volume, assuming that the pressure remains constant.

9.6    The initial temperature of the air is found to be 75 °F when the volume is found 2 ft$^3$. The temperature is then raised to 115 °F. Calculate the volume of the air after being heated.

9.7    The initial temperature of the air is found to be 55 °F when the volume is found 1.5 ft$^3$. The temperature is then raised and the volume is found to be 6 ft$^3$. Calculate the final temperature after being heated.

9.8    A gas of 25 bar (gauge) is stored in a 1.5 m$^3$ volume cylinder at a temperature of 75 $^{°C}$. The piston of the cylinder compressed the volume to 0.5 m$^3$ while the gas is heated to 112 °C. Calculate the final gauge pressure.

9.9    Air is initially stored at 0 psig and 75 °F to a volume of 12 ft$^3$. Then the volume is reduced to 1.5 ft$^3$ and it is found that the temperature of the air increases to 100 °F. Calculate the final pressure of the air.

9.10   Air is initially stored at 0 psig and 75 °F to a volume of 12 ft$^3$. Then the volume is reduced to 1.5 ft$^3$ by using the 80 psi pressure. Calculate the temperature of the air.

9.11   In an adiabatic process, the air is compressed from 8.5 ft$^3$ to 2.5 ft$^3$ at 75 °F and atmospheric pressure. Calculate the final pressure and temperature of the gas.

9.12   In an adiabatic process, the air is compressed from 5.5 m$^3$ to 1.2 m$^3$ at 45 °C and atmospheric pressure. Determine the final pressure and temperature of the gas.

9.13   A compressor delivers 45 scfm of air at 85 psig. The overall efficiency of a compressor is 70%. Determine the theoretical power and apparent power.

9.14   The compressor delivers 6 standard m$^3$/min at 625 kPa. The overall efficiency of a compressor is 85%. Calculate the theoretical power and apparent power.

9.15   A pneumatic system demands 55 scfm and the compressor delivers 30 scfm. The receiver runs for 4 min and the pressure reduces from 60 to 90 psi. Determine the receiver size.

9.16   A pneumatic system demands 65 m³/min and the compressor delivers 35 m³/min. The receiver runs for 8 min and the pressure reduces from 125 to 60 kPa. Calculate the receiver size.

9.17   An air receiver with a size of 6 scfm is charged to 120 psi. The compressor delivers 3 scfm at 80 psi before going to shut down. Calculate the time.

9.18   An air receiver delivers the mean flow rate of 6 m³/min at 500 kPa for 5 min and then the compressor shuts down. Determine the size of the receiver if it is charged at 750 kPa.

9.19   The inside and outside diameters of suction cups are found to be 1.5 cm and 4.5 cm, respectively. Calculate the lifting force for—12 kPa = 12 kPa suction = 12 kPa vacuum.

9.20   A pneumatic system lifts a workpiece using a total volume of 6 ft³ inside the suction cup and associated pipeline up to the vacuum pump. The atmospheric and the suction pressures are 14.7 psia and 6 psia. Calculate the required time to obtain the required level of vacuum pressure when the pump produces a flow rate of 6 scfm.

# References

1. A. Esposito, *Fluid Power with Applications*, 7th edn. (Pearson Education Limited, Edinburgh Gate, Essex, England, 2014)
2. K. Subramanya, *Fluid Mechanics and Hydraulic Machines* (Tata McGraw Hill Education Private Limited, New Delhi, 2011)
3. F. Don Norvelle, *Fluid Power Technology*, 1st edn. (West Legal Studies is an imprint of Delmar, a division of Thomson Learning, New York, 1995)

# Chapter 10
# Pneumatic System Components

## 10.1 Introduction

The different types of pneumatic system components are usually used to operate and control any pneumatic system in a fluid power system. These are manifold, fluid conditioning units usually known as filter, regulator and lubricating (FRL), fluid control valve, directional control valve, connecting tubes and cylinders. These components are discussed in this chapter.

## 10.2 Pneumatic Cylinders

The pneumatic cylinders are very similar to hydraulic cylinders. The pneumatic cylinders are muscles of the pneumatic systems as they are used to move, lift and hold the objects based on the respective applications. The cylinders are the devices that produce the force with the help of compressed gas. A pneumatic cylinder with two eyes and ports is shown in Fig. 10.1 [1].

The pneumatic cylinder converts the stored energy in the compressed gas into linear motion. Normally, the pressure is applied to the piston of the cylinder that moves in the linear direction. The mechanical generated by the cylinder is calculated by the fundamentals of Pascal's law that is the force is equal to the product of the compressed gas pressure and the area of the cylinder.

Each cylinder has two ends namely cap or blind end and rod end. The closed end of the cylinder is known as cap-end or dead-end or blind-end. Whereas the open end of the cylinder is known as head-end or rod-end or head-end and the cylinder rod normally comes through the rod end. The proper sealing (Dynamic and static) is used to make the cylinder leakproof. Static seals are used inside the barrel and dynamic seals are used with the piston. The wiper seals are also used at the rod-end to prevent the dirt, water and other unwanted contaminants from entering into the cylinder during retraction. The tube part of the cylinder is known as the barrel that provides

Md. A. Salam, *Fundamentals of Pneumatics and Hydraulics*,
https://doi.org/10.1007/978-981-19-0855-2_10

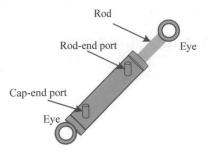

**Fig. 10.1**  A sample single-acting cylinder

the desired bore diameter during extension and retraction with zero leakings. The barrel also provides mechanical strength and protection during operation at a specific pressure. The main components of a real cylinder are shown in (Figs. 10.2 and 10.3).

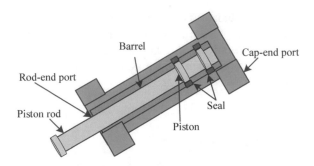

**Fig. 10.2**  Construction of a cylinder

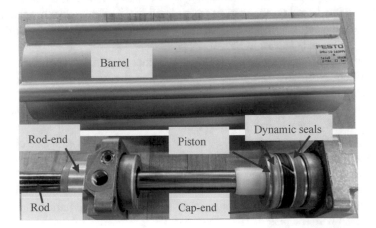

**Fig. 10.3**  Construction of a cylinder

Usually, the pneumatic cylinders operate at a much lower pressure than the hydraulic cylinders. Therefore, the pneumatic cylinders are made of aluminum. Based on applications, the pneumatic cylinders are classified as low-duty, medium-duty, and heavy-duty cylinders. In general, based on cylinder action, the pneumatic cylinders are classified as single-acting, double-acting (single-rod and double-rod), and telescopic or multistage.

### 10.2.1  Sigle-Acting Cylinder

There is one connection port in a single-acting cylinder where the compressed air supply is connected. A single-acting cylinder with and without spring is shown in Fig. 10.4. A compressed air supply is connected to a piston side port of a spring control single-acting cylinder as shown in Fig. 10.5. As a result, the pressure of the compressed air will act on the piston of the cylinder and pushes the piston. If the force of the compressed air is higher than the spring force, then the piston rod will be fully outstroke.

This position of the cylinder is known as an extension as shown in Fig. 10.5. When the air supply is stopped, then the spring pressure will be higher than the supply pressure. As a result, the air from the cylinder will go out and it will be at a fully instroke position. This position of the cylinder is known as the retraction position as shown in Fig. 10.6.

The simulation of a single-acting cylinder is carried out using an air supply and a 3/2 normally closed directional control valve. The extension and retraction of a single-acting cylinder are shown in Fig. 10.7.

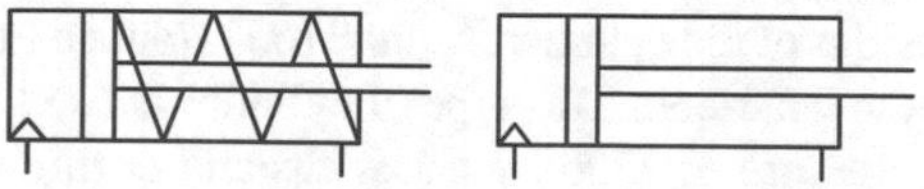

**Fig. 10.4**  A single-acting cylinder with and without spring

**Fig. 10.5**  A single-acting cylinder with extension

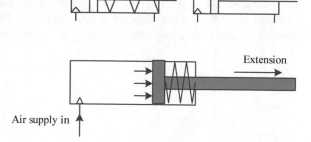

Extension

Air supply in

**Fig. 10.6**  A single-acting cylinder with retraction

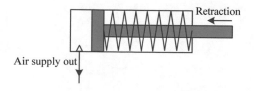

Retraction

Air supply out

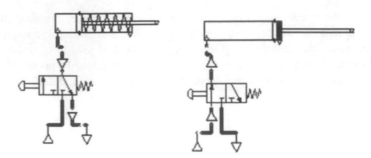

**Fig. 10.7**   A single-acting cylinder with retraction and attraction

## 10.2.2   Double-Acting Cylinder

A double-acting cylinder has two connection ports as it does not have a spring. One connection port is near the cap-end terminal and the other port is near the rod-end terminal. During outstroke, the compressed air supply is connected to the cap-end port and the exhaust is connected to the rod-end port. In this case, the cylinder is fully outside of the barrel and this position is known as an extension of the cylinder as shown in Fig. 10.8. Whereas in the return motion, the compressed air supply is connected to the rod-end port and the exhaust is connected to the cap-end. This position of the cylinder is known as retraction as shown in Fig. 10.9. The double-acting cylinder is used in a pneumatic system where a large magnitude of the force is required to move the workpiece [2].

There are some special applications where a double-acting cylinder with ended double rods is used. An example of such a load is simultaneous load. A given flow rate produces the same extension and retraction with equal area and volume on both sides of the cylinder. A small rod is placed in the center position of the piston to create the dead-end. This type of cylinder is used in a hydraulic system and the maximum pressure of 2500 psi can withstand during operation. The construction and graphic symbol of a double-acting ending with double rods are shown in Fig. 10.10a, b,

**Fig. 10.8**   A double-acting cylinder with an extended position

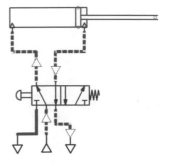

**Fig. 10.9** A double-acting cylinder with the retracted position

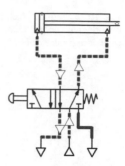

**Fig. 10.10** A double-acting cylinder with ended double rods and graphic symbol

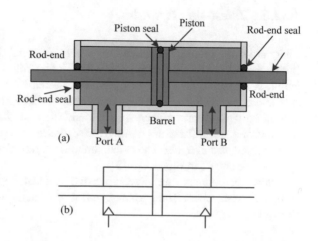

respectively. The extension and retraction of double-acting double rods are shown in Figs. 10.11 and 10.12, respectively.

**Fig. 10.11** A double-acting cylinder with ended double rods during extension

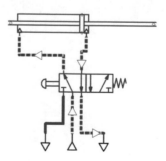

**Fig. 10.12**  A double-acting
cylinder with ended double
rods during retraction

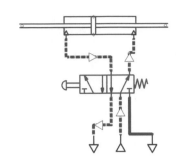

### 10.2.3  Telescopic Cylinder

There are some industrial applications where a special design cylinder is required.
In this type of application, a cylinder with a long stroke length and a short retracted
length is required. This type of operation offered by a special cylinder is known
as a telescopic cylinder. This type of cylinder is designed for high pressure that is
around 2700 psi to carry a heavyweight load. There are many places where telescopic
cylinders are used. These are dump trucks, amusement parks, during a long vehicle
carrying new vehicles, etc. The extension and the retraction positions of a telescopic
cylinder are shown in Fig. 10.13 [3].

Telescopic cylinders are more expensive than standard cylinders because of their
complex construction. The graphic symbol of a single-acting telescopic cylinder is
shown in Fig. 10.14.

**Fig. 10.13**  An extension
and a retracted position of a
telescopic cylinder

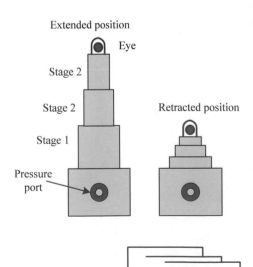

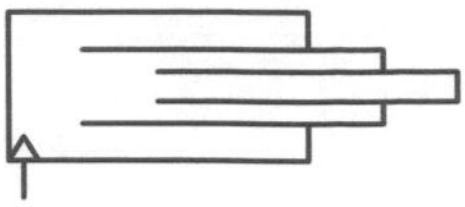

**Fig. 10.14**  A graphic
symbol of a single-acting
telescopic cylinder

**Fig. 10.15** A graphic symbol of a double-acting telescopic cylinder

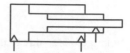

In Automation Studio software, the telescopic cylinder is only found under the hydraulic circuit. In this software, both double-acting and single-acting telescopic cylinders are available and can be designed with different stages. However, in MS Visio, the symbols of single-acting and double-acting telescopic cylinders are available. The graphic symbol of a double-acting telescopic cylinder is shown in Fig. 10.15.

The hydraulic control valves, pressure relief valve, pump, reservoir, and tank are used to understand the operation of a telescopic cylinder. Under normal simulation, pushes the lever of the directional control valve and the piston of the telescopic cylinder will come out from the original position and extend as shown in Fig. 10.13.

## 10.3 Formulas for Pneumatic Cylinder

In a practical application, the pneumatic cylinder is extended and retracted faster due to the compressibility of air. In a pneumatic cylinder, lower pressure is used so that it produces smaller magnitudes of the forces during extension and retraction. Pascal's law is applied to determine the magnitude of the extension force and retraction force. A double-acting cylinder with extension and retraction forces is shown in Fig. 10.16 [4].

The force during extension is expressed as,

$$F_e = pA_p \tag{10.1}$$

where the area due to the piston is expressed as,

$$A_p = \pi \frac{D^2}{4} \tag{10.2}$$

The force during retraction is expressed as,

**Fig. 10.16** A double-acting cylinder with extension and retraction force

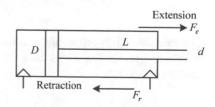

$$F_r = p(A_p - A_r)$$                                                                    (10.3)

where the area due to the rod is expressed as,

$$A_r = \pi \frac{d^2}{4}$$                                                                    (10.4)

where,

D    is the diameter of the piston in in., m,
d    is the diameter of the piston rod in in., m,
$F_e$    is the extension force in lb, N,
$F_r$    is the retraction force in lb, N,
$A_p$    is the area of the piston in in.$^2$, m$^2$,
$A_r$    is the area of the piston in in.$^2$, m$^2$.

**Example 10.1** The piston and rod diameters of a pneumatic cylinder are 2 in. and 1.5 in., respectively. Calculate the extension and retraction forces if the system pressure is 50 psi.

**Solution**
The area of the piston is calculated as,

$$A_p = \frac{\pi D^2}{4} = \frac{\pi \times 2^2}{4} = 3.14 \, \text{in.}^2$$                                                                    (10.5)

The area of the rod is calculated as,

$$A_p = \frac{\pi D^2}{4} = \frac{\pi \times 1.5^2}{4} = 1.77 \, \text{in.}^2$$                                                                    (10.6)

The force during extension is calculated as,

$$F_e = pA_p = 50 \times 3.14 = 157 \, \text{lb}$$                                                                    (10.7)

The force during retraction is calculated as,

$$F_e = F_r = p(A_p - A_r) = 50 \times (3.14 - 1.77) = 68.5 \, \text{lb}$$                                                                    (10.8)

**Example 10.2** The piston and rod diameters of a pneumatic cylinder are 50 mm and 40 mm, respectively. Calculate the extension and retraction forces if the system pressure is 1500 Pa.

**Solution**
The area of the piston is calculated as,

$$A_p = \frac{\pi D^2}{4} = \frac{\pi \times 0.050^2}{4} = 0.0019 \, \text{m}^2 \qquad (10.9)$$

The area of the rod is calculated as,

$$A_p = \frac{\pi D^2}{4} = \frac{\pi \times 40^2}{4} = 0.0013 \, \text{m}^2 \qquad (10.10)$$

The force during extension is calculated as,

$$F_e = pA_p = 1500 \times 0.0019 = 2.85 \, \text{N} \qquad (10.11)$$

The force during retraction is calculated as,

$$F_e = F_r = p(A_p - A_r) = 1500 \times (0.0019 - 0.0013) = 0.9 \, \text{N} \qquad (10.12)$$

**Practice Problem 10.1**
The piston and rod diameters of a pneumatic cylinder are 2.5 in. and 1.8 in., respectively. Calculate the extension and retraction forces if the system pressure is 40 psi.

**Practice Problem 10.2**
The piston and rod diameters of a pneumatic cylinder are 65 mm and 45 mm, respectively. Calculate the extension and retraction forces if the system pressure is 15000 Pa.

## 10.4  Flow Coefficient of a Pneumatic Valve

The proper design of a pneumatic system is important to save money. There is an important parameter such as flow coefficient or flow factor that needs to be considered for designing the pneumatic system. The design expressions for $C_v$ for a pneumatic valve in FPS and MKS units are expressed as,

$$C_v = \frac{Q}{22.48} \sqrt{\frac{S_g \times T}{(p_1 - p_2)p_2}} \qquad (10.13)$$

$$C_v = \frac{Q}{68.7} \sqrt{\frac{S_g \times T}{(p_1 - p_2)p_2}} \qquad (10.14)$$

$$\frac{p_2}{p_1} \geq 0.5 \quad \text{maximum flow} \tag{10.15}$$

$$\frac{p_2}{p_1} < 0.5 \quad \text{chocked flow} \tag{10.16}$$

where,

Q     is the flow rate in standard cubic feet per min (SCFM), LPM.
$p_1$     is the inlet absolute pressure (gauge pressure + 14.7),
$p_2$     is the outlet absolute pressure (gauge pressure + 14.7),
$S_g$     is the specific gravity (for air, $S_g = 1$).
T     is the absolute temperature (°F + 460) °R, (°C + 273) °K,

**Example 10.3** A directional control valve is used to control an airpower machine. This machine consumes air 75 SCFM at 80 psig. The pressure drops across the valve is 10 psig. Calculate the valve flow factor if the air temperature is 70°F.

**Solution**
The output pressure in absolute value is calculated as,

$$p_2 = 80 + 14.7 = 94.7 \, \text{psia} \tag{10.17}$$

The temperature in °K is calculated as,

$$T = 70 + 460 = 530°\text{K} \tag{10.18}$$

The pressure drop in psia value is calculated as,

$$\Delta p = p_1 - p_2 = 10 + 14.7 = 24.7 \, \text{psia} \tag{10.19}$$

The flow factor is calculated as,

$$C_v = \frac{Q}{22.48} \sqrt{\frac{S_g \times T}{(p_1 - p_2)p_2}} = \frac{75}{22.48} \sqrt{\frac{1 \times 530}{24.7 \times 97.4}} = 1.56 \tag{10.20}$$

According to relevant standards, the value of the flow factor of 1.13 or higher provides the information that the pressure drop across the valve will be no more than 5 psi.

**Example 10.4** The value of a flow factor of 90 psig (line pressure) pneumatic system is found to be 1.14. The pressure drop across the valve is found to be 4.5 psig. Determine the flow rate if the air temperature is 70°F.

**Solution**

The output pressure is calculated as,

$$p_2 = 90 - 4.5 = 85.5\,\text{psig} = 85.5 + 14.7 = 100.2\,\text{psia} \tag{10.21}$$

The pressure drop in psia value is calculated as,

$$\Delta p = 4.5 + 14.7 = 19.2\,\text{psia} \tag{10.22}$$

The temperature in °K is calculated as,

$$T = 70 + 460 = 530°\text{K} \tag{10.23}$$

$$Q = 22.48C_v\sqrt{\frac{\Delta p \times p_2}{T}} = 22.48 \times 1.14\sqrt{\frac{19.2 \times 100.2}{530}} = 48.82\,\text{scfm} \tag{10.24}$$

**Practice Problem 10.3**

A directional control valve is used to control an airpower machine. This machine consumes air 65 SCFM at 70 psig. The pressure drops across the valve is 8 psig. Calculate the valve flow factor if the air temperature is 65°F.

**Practice Problem 10.4**

The value of a flow factor of 100 psig (line pressure) pneumatic system is found to be 1.14. The pressure drop across the valve is found to be 9.5 psig. Find the flow rate if the air temperature is 80°F.

## 10.5  Numbering of Pneumatic Valves

Pneumatic directional control valves (DCV) are used in different types of applications. The connection ports and position of any directional control valves need to identify before using them in the circuit. The number of positions of DCVof a valve refers to the number of blocks. Whereas the number of ports refers to the number of connection points. The DCV is usually identified based on the functions, actuation methods, number of connection ports, and positions. ISO standards such as ISO 1219–1: 2006 specified the graphic symbols and circuit diagram, ISO 1219–2:1995 established the guidelines for drawing diagrams, ISO 5599 standardized the port designation of fluid power systems components. According to ISO 5599, ports are levelled using the number systems. The old and ISO numbering systems of DCV are shown in Table 10.1.

**Table 10.1** Different levels of vacuum

| Port name | Old numbering | ISO numbering system |
| --- | --- | --- |
| Pressure port | P | 1 |
| Auxiliary or working ports | A, B | 2, 4 |
| Exhaust ports | R, S | 5, 3 |

**Fig. 10.17** A graphic symbol of a pneumatic pressure relief valve

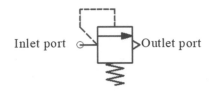

Inlet port  ⊶                 ▷ Outlet port

## 10.6   Pneumatic Pressure Relief Valve

A pneumatic pressure relief valve is used in a system from overpressure for the safety of other relevant equipment and personnel. This valve is spring-loaded that opens when the preset maximum pressure is reached to prevent the further increase in pressure. If the pressure exceeds the amount allowed by the pressure relief valve, the pressure relief valve will automatically open and release the excess air until the pressure is reduced. The graphic symbol of the pneumatic pressure relief valve is shown in Fig. 10.17.

The relief valve blocks the fluid from flowing from port 1 to port 2 when the pressure is insufficient at the inlet port. However, when the pressure is sufficient that is greater or equal to the cracking pressure, the fluid flows from the inlet port to the outlet port.

## 10.7   Pneumatic Valves with Actuations

Different types of pneumatic valves are shown in Fig. 10.18. According to the requirement, these valves are used in the pneumatic circuits to control the operation of the cylinders.

## 10.8   Filter Regulator Lubricator

The compressed air becomes hot, dirty and wet when comes out from the compressor. This condition of the air shortens the downstream pneumatic equipment such as directional control valve, cylinder as well as other relevant tools. A combined device

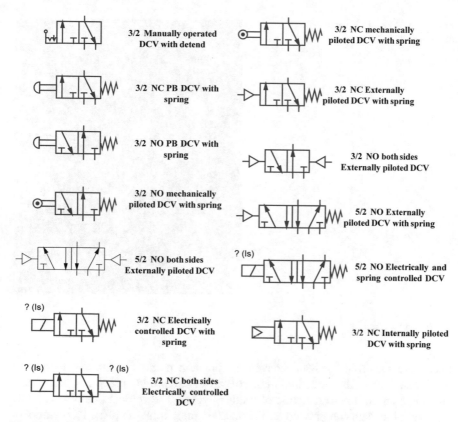

**Fig. 10.18**  Symbols of pneumatic valves

named FRL comes in to provide better working conditions. An FRL combines the airline filter, pressure regulator, and lubricator into one component in the pneumatic circuit to provide the maximum working condition of the air compressor. The graphical symbol of FRL is shown in Fig. 10.19.

In a pneumatic trainer kit, FRL is used to provide pressurized air to the manifold as shown in Fig. 10.20. Then the air is circuited to the required directional control valve by using plastic tubes. This FRL has a shut-off valve to open and close the pressurized air. In addition to that, there is a black colour knob to vary the air pressure.

**Fig. 10.19**  Symbol FRL

Inlet port  Outlet port

**Fig. 10.20** A real FRL used
in trainer kit

## 10.9   Manifold

Pneumatic manifolds provide convenient junction points for the distribution of
compressed air to the directional control valves. Manifolds provide a convenient
junction point for the distribution of fluids or gases. The vertical line of Fig. 10.21
represents the main compressed airline that is coming from the FRL. The symbol of
the manifold that is used in the pneumatic trainer kit is shown in Fig. 10.22.

**Exercise Problems**

10.1   The piston and rod diameters of a pneumatic cylinder are 2.2 in and 1.6
       in, respectively. Calculate the extension and retraction forces if the system
       pressure is 45 psi.

**Fig. 10.21** Eight points of
manifold

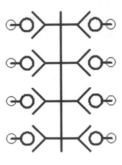

**Fig. 10.22** A real manifold
used in the trainer kit

10.2　The piston and rod diameters of a pneumatic cylinder are 60 mm and 50 mm, respectively. Calculate the extension and retraction forces if the system pressure is 18000 Pa.

10.3　A directional control valve is used to control an airpower machine. This machine consumes air 55 SCFM at 86 psig. The pressure drops across the valve is 10 psig. Calculate the valve flow factor if the air temperature is 75°F.

10.4　The value of a flow factor of 110 psig (line pressure) pneumatic system is found to be 1.16. The pressure drop across the valve is found to be 10.5 psig. Find the flow rate if the air temperature is 85°F.

# References

1. A. Esposito, *Fluid Power with Applications*, Seventh Edn. (Pearson Education Limited, Edinburgh Gate, Essex, England, 2014)
2. K. Subramanya, *Fluid Mechanics and Hydraulic Machines*, (Tata McGraw Hill Education Private Limited, New Delhi, 2011)
3. F. Don Norvelle, *Fluid Power Technology*, First Edn. (West Legal Studies is an imprint of Delmar, a division of Thomson Learning, New York, 1995)
4. J.L. Johnson, *Introduction to Fluid Power*, First Edn. (Delmar, a division of Thomson Learning, The United States of America, 2002), pp. 1–502

# Chapter 11
# Electrical Devices and Control of Cylinders

## 11.1  Introduction

An electro-pneumatic control is a combination of pneumatic and electrical technologies. In this control, 24 V AC or DC source is used. However, compressed air is used as the main working medium. The directional control valve is operated by a solenoid actuation. The resetting of the directional control valve is obtained by using a spring or another solenoid. The control activities are obtained by the combination of the electromagnetic relay, push-button switches and magnetic contactors. In this chapter, the details of these electrical components and different electro-pneumatic circuits are discussed.

## 11.2  Different Types of Switches

The electrical components which can make or break electrical circuits manually or automatically are known as switches. The switch has two states such as normally open (NO) and normally close (NC). In a switch, the pole and throw are two important items. The pole represents the number of circuits controlled by a switch. Whereas the throw refers to the number of output connection points that can make the switch to the input. In general, the pole is the number of moving portions of a switch and the throw is the number of different connections of a switch. Based on the pole and throw, the electrical switches are classified as single pole single throw (SPST), single pole double throw (SPDT), double pole single throw (DPST) and double pole double throw (DPDT). An SPST switch performs the basic ON/OFF function of a single circuit with two terminals shown in Fig. 11.1a [1].

An SPDT switch connects the input to the two output circuits as shown in Fig. 11.1b. A DPST switch is often used to connect two inputs to the two output circuits and it has four different terminals as shown in Fig. 11.1c. A DPDT switch

**Fig. 11.1** Different types of switches

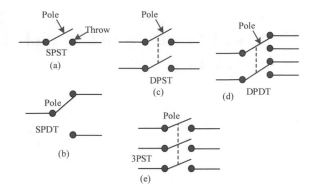

has six terminals and connects two independent output circuits as shown in Fig. 11.1d. A 3PST switch connects three inputs with three outputs as shown in Fig. 11.1e.

## 11.3   Push-Button Switch

The push-button switches are electrical switches that are activated by pressing (external force). According to operation, it is classified as start push-button and stop push-button switch. The push-button switches are used in a control circuit to start, stop, forward, and reverse rotation. The start and stop push-button switches have two pairs of contacts. The start push-button switch has two-pair of contacts that is normally open. In a deactivated state, it is connected when the button of the push-button is pressed as shown in Fig. 11.2.

While the stop-button switch has also two pairs of contacts that are normally closed. In an activated state, it is disconnected or opened when the button is pressed as shown in Fig. 11.3. There are different colours of push-button switches are available in the market. These are red, green, black, yellow, blue, white, etc. The red colour is used to stop the push-button switch. Whereas green, black and other colours are used to start the push-button switch.

**Fig. 11.2** A start push-button switch and symbol

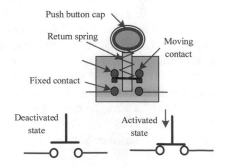

**Fig. 11.3**  A stop push-button switch and symbol

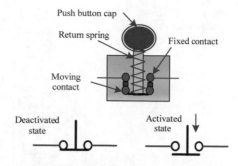

**Fig. 11.4**  A start and stop push-button switches

In a lathe machine, both green and black colours start push-button switches and one red colour stops push-button switches are used to operate. However, the combination of green colour and red colour push-button switches are used to operate a three-phase induction motor in the industry.

The real black and red colour push-button switches are shown in Fig. 11.4.

## 11.4  Limit Switch

The limit switch is an electromechanical device that detects the presence or absence of an object with a contact or non-contact sensing device. The sensors then produce an electrical output signal that can be used to control equipment or process. A limit switch detects the presence of a moving object when it reaches the desired location. The limit switch is used in the refrigerator to identify if the door is closed or open. The different configurations of the limit switch are shown in Fig. 11.5

The limit switch is an electromechanical device that consists of a mechanical actuator that is linked to a series of electrical contacts. In other words, a limit switch is an electromechanical device operated by a physical force applied to it by an object. In a garage door, the limit switch is used to stop the movement of the door when it reaches its fully opened position. It is also used in the car door and plane door opening and closing activities.

**Fig. 11.5** Different
configurations of the limit
switch

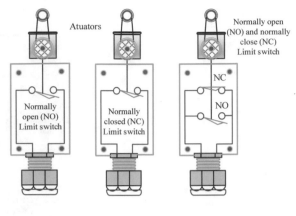

**Fig. 11.6** Configurations
and symbols of a pressure
switch

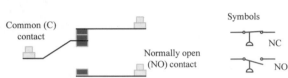

## 11.5   Pressure Switch

The pressure switches open or close their contacts based on the system pressure. The switch makes electrical contact when a certain preset pressure is reached to its input is known as a pressure switch. In other words, the pressure switch is used to measure and maintain the pressure or level of liquid or gas media. These switches have three electrical terminals namely a common (C), NC (normally closed) and NO (normally open). Only two terminals are used during wiring (common and NC or NO). The physical configuration and the symbols of pressure switches are shown in Fig. 11.6.

Pressure switches are widely used in residential and industrial applications such as HVAC, pumps, furnaces etc.

## 11.6   Level Switch

The switch that detects the level of liquid or solid in a container is known as a level switch. The level switch is often known as a float switch as it floats on a liquid container. A float switch is a mechanical switch that floats on the top of a liquid surface and moves vertically as the liquid level goes up or down. When it is used as a level sensing element, the motion of which actuates the contacts to open or close. The normally open and normally closed level switch is shown in Fig. 11.11. These are similar to pressure switches except a sphere ball is used instead of a semi-sphere ball (Fig. 11.7).

**Fig. 11.7**  Symbols of a level
switch

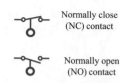

Normally close
(NC) contact

Normally open
(NO) contact

The level switch is used in a washing machine as a level sensor is to know when water needs to be pumped into the drum and when water needs to be pumped out. The various wash cycles mainly depend on various water levels in the drum which mainly used the float switch. The float switch has a magnet that is free to move up and down along the shaft as the water level changes. The float switch is used to trip alarms, engage relays, control pumps and valves. It is also used for heater protection. There are two types of float switches such as stem-mounted float switch and cable-suspended float switch.

## 11.7   Proximity Switch

In industrial applications, a proximity switch is used to count or record moving objects or work-piece on machines or conveyors. A switch that is used to detect the proximity or closeness of an object is known as a proximity switch. This switch is a non-contact sensor that uses magnetic, electricity, optical means to sense the proximity of objects. This switch is incorporated with power electronics and it is enclosed by a diamond shape that indicates a powered or active device as shown in Fig. 11.8.

The proximity switch is classified as inductive, capacitive and optical types. In an inductive proximity switch, an oscillator with LC resonant circuit is used to create a high-frequency oscillation magnetic field across the active switching zone and detects accordingly. The symbol of the inductive proximity sensor is shown in Fig. 11.9. Here, Fe represents an inductive, PNP and NPN represents the bipolar transistor.

Whereas in a capacitive proximity switch, an oscillator with RC resonant circuit is used to create the magnetic field. When the object is at a preset distance, then the capacitive sensor uses the variation of capacitance to detect it. The change of oscillation (increase or decrease) is detected by a threshold circuit that changes

**Fig. 11.8**  Symbols of the
proximity switch

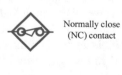

Normally close
(NC) contact

Normally open
(NO) contact

**Fig. 11.9** Symbols of the inductive proximity switch

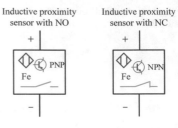

**Fig. 11.10** Symbols of the capacitive proximity switch

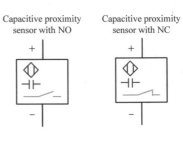

**Fig. 11.11** Symbols of the optical fibre proximity switch

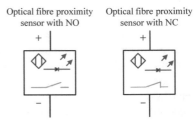

the sensor output. The symbols for the capacitive proximity sensors are shown in Fig. 11.10. This type of sensor is used to monitor pressure, fluid level, motion etc.

A sensor measures a physical quantity based on its modulation on the intensity, spectrum, phase, or polarization of light travelling through an optical fibre is known as an optical fibre sensor. This type of sensor is used to monitor temperature, pressure, vibration, rotation etc. The symbol of the optical fibre sensor is shown in Fig. 11.11. The photodiode (LET) detects the light generated. This type of sensor detects the object when it interrupts a light beam. It can sense any object remotely within the 10-m range.

One of the important applications of proximity sensors in the plane landing gear. The landing gear control unit reads the proximity sensors and the flight control system receives the signals indicating deployment of ground spoilers by the landing gear control unit. The ground spoiler is used in a plane to maximize the wheel brake efficiency by spoiling or dumping the lift generated by the wing and thus imposing the full weight of the aircraft onto the landing gear.

**Fig. 11.12** Symbols of the
temperature switch

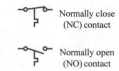

Normally close
(NC) contact

Normally open
(NO) contact

## 11.8  Temperature Switch

A switch that automatically senses a change in the temperature of an object and opens or closes an electrical switch is known as a temperature switch. The temperature switches use bi-metallic strips as a sensing element and its normally closed and open symbols are shown in Fig. 11.12.

These temperature sensors are used in a motor engine to monitor engine conditions and bearing temperatures. These are also used in HVAC (heating ventilation and air conditioning) systems for controlling the temperature of the university building, hospital building, shopping malls, and industrial and commercial buildings.

## 11.9  Flow Switch

A switch that detects the flow rate of fluid through a pipe is known as a flow switch. The flow switches are classified as paddles, thermal, piezo and shuttle or piston. Paddles are often used as a flow-sensing element that opens or closes electrical contacts. The paddle type flow switch is used in boiler flow, air conditioning, controlling dampers, motor protection etc. In a thermal type flow switch, the thermals are often used as a flow-sensing element that opens or closes electrical contacts. This type of flow switch is used in water heaters, chillers and liquid transfer systems. The piezo-type flow switch uses the piezoelectric effect to measure the flow rate in a system. It is used to monitor bulk flow rate in a pipe, spray systems, rotary drum and food processing system. The shuttle or piston-type flow switch provides electrical contact based on the flow rate. The shuttle is displaced as the liquid flow increases to the actuation area and it returns to the original position as the liquid flow decreases. The normally open (NO) and normally closed (NC) symbols are shown in Fig. 11.13.

These flow switches send an electrical signal to electronic controllers in case the flow rate is too fast or slow.

**Fig. 11.13** Symbols of the
flow switch

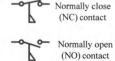

Normally close
(NC) contact

Normally open
(NO) contact

## 11.10   Solenoid

A solenoid refers to coils wrapped around a core that creates uniform magnetic fields when current passes through it as shown in Fig. 11.14. The solenoid also refers to different types of transducers that convert electrical energy into a linear motion (mechanical energy). Therefore, the solenoid valve is an electromechanical device that actuates either a pneumatic valve or hydraulic valve.

A solenoid valve consists of the solenoid, valve body, coil, iron core, plunger, shading rings. When AC or DC supply connects to the solenoid, the current flows through the coil that creates the magnetic field. This magnetic field is expressed as,

$$F = BLI \tag{11.1}$$

where,

$B$   is the magnetic field in Wb/m$^2$,
$L$   is the length of the coil or conductor in m,
$I$   is the current in A.

This magnetic force creates an attraction that opens the orifice and allows the flow through the valve. If the valve is open initially, then it blocks the orifice and stops the flow through the valve. The shading rings prevent vibration and noise that generates in the coils.

Again, the current flows through a coil that creates the internal and external magnetic fields as shown in Fig. 11.15.

The magnetic field strength depends on the number of turns of the coil and the magnitude of the current that flows through the coil. Mathematically, it is expressed as,

$$H = \frac{\Im}{l} = \frac{NI}{l} \quad \text{At/m} \tag{11.2}$$

The solenoid valves are available in different AC and DC voltages. When an AC voltage is used to energize the solenoid coil, then it is known as an AC solenoid. If a solenoid coil is energized by a DC voltage then it is known as a DC solenoid. The different values of energized DC voltages are 3 V, 6 V, 12 V, 24 V, and 48 V DC. The

**Fig. 11.14**  Coils with magnetic fields

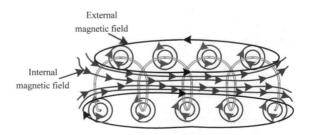

External magnetic field

Internal magnetic field

**Fig. 11.15** A current with magnetic fields

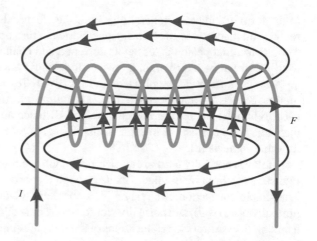

**Fig. 11.16** Push and pull types solenoid

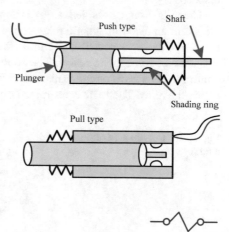

**Fig. 11.17** Graphical symbol of a solenoid

AC voltages at 50 Hz are 24 V, 110 V, 120 V, and 220 V. The AC voltages at 60 Hz are 24 V, 110 V, and 208 V. Several types of solenoids are used practically. However, the push and pull types are shown in Fig. 11.16.

The graphical symbol of a solenoid that is used in the fluid power is shown in Fig. 11.17.

## 11.11 Relay

A relay is an electromagnetic or solid-state device that controls a specific device or circuit under certain conditions. In other words, a relay is a smart device that receives input signals, compares them to the set values, and provides an output signal. The input signals may be current, voltage, or temperature. Whereas the output signals are in the form of indicator lights. In general, the electrical supply is connected to the relay's coil that creates an electromagnet which in turn opens and closes the contacts from their initial position. A relay is designed and connected to an electrical circuit to trip a circuit breaker within a few seconds under abnormal conditions. In 1835, an American scientist Joseph Henry, Jr. (1797–1878) invented electromagnetism, motors, generators, and the telegraph. At that time, electromagnetic relays were first used to transfer the telegraph signal to long-distance amplifiers. A typical electromagnetic relay consists of the coil, armature, core, yoke, pivot, common, fixed, and movable contacts [2].

A relay is normally used in any electrical control circuit to control the three-phase induction motor or other related loads through its normally open or closed contacts. The simple connection of a relay with a coil, supply voltage, spring, common (C), normally open (NO), and normally closed (NC) connections are shown in Fig. 11.18. The simple base or socket connections of the relay are shown in Figs. 11.19 and 11.20, respectively.

The relays are classified as electromagnetic, solid-state, and microprocessor relays. In an electromagnetic relay, the armature is activated by energizing the coil to open or close an electrical circuit. In the late 1960s, solid-state relays appeared to replace electromagnetic relays. The solid-state relays use semiconductor output to control the electrical circuit. This output is optically coupled to an LED light source inside the relay. The LED is energized by a low voltage DC power supply which in

**Fig. 11.18** Construction of a relay

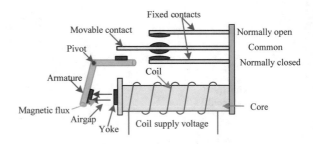

**Fig. 11.19** A simple relay connection

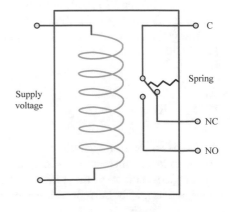

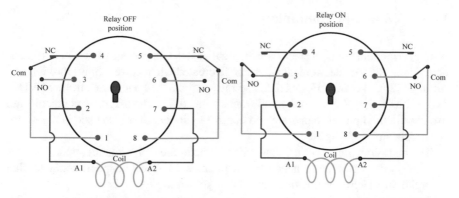

**Fig. 11.20** A relay connection with a socket

turn is on the relay. The solid-state relays do not have moving parts, speed is very high, and the reset time is less than the electromagnetic relay. The solid-state relay could also be set precisely and it has less maintenance cost.

The microprocessor relays provide many functions compared to electromagnetic and solid-state relays. This relay has two logic states (1 or 0) namely ON state or high level (1) and OFF state (0) or low level. Its function depends on the AND, OR, and NOT logic gates. The microprocessor relays are used in power system protection that utilizes software-based numerical technique measurements. A relay socket and top part are shown in Fig. 11.21. A relay changes its deactivated state into an activated state when the coil is energized by AC/DC voltage.

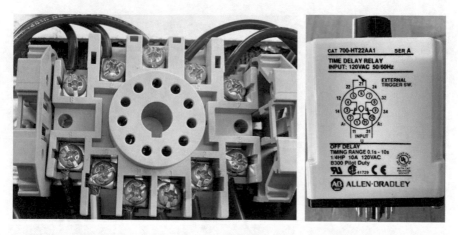

**Fig. 11.21** A relay socket and top part

## 11.12  Magnetic Contactor

A magnetic contactor is an electromechanical switch that operates due to electromagnetic induction. An electromagnetic force is created when current flows through coil A1 and A2 as shown in Fig. 11.22. The rating of coil voltage varies from 24 VDC, 110 V, 120 V, 220 V AC based on relevant applications. The magnetic contactors are used in the fluid power to operate and control cylinders. It is also used in the motor control, heater, lighting etc.

The main components of the magnetic contactor are coil, armature, main contacts (normally open), auxiliary contacts (open and closed). A real magnetic contactor that is used in an electro-pneumatic trainer kit is shown in Fig. 11.23.

The main and auxiliary contacts along with the coil are shown in Fig. 11.24. The auxiliary closed contacts are numbered 1 is the line side and 2 is the load side. Whereas the auxiliary open contacts are numbered 3 is the line side and 4 is the load side. When the coil terminals of the contactor are connected to a source, then the current flows through the coil creates the magnetic force ($F = BIl$). This magnetic force attracts the armature which in turn changes the contact states. That means normally open contacts will be closed and vice versa. If the coil is disconnected from the supply, then the coil will be de-energized. As a result, the armature of the magnetic contactor will come back to the initial position which means the normally open contacts will remain open and the normally closed contacts will remain closed.

There is a difference between the magnetic contactor and relay. The relay has the coil, normally open and close contacts. The relay operates and changes the states based on the specific time setting. The relay has a smaller number of contacts than the magnetic contactor.

**Fig. 11.22**  Construction of a magnetic contactor

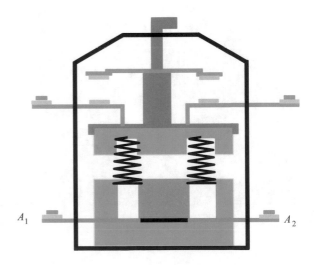

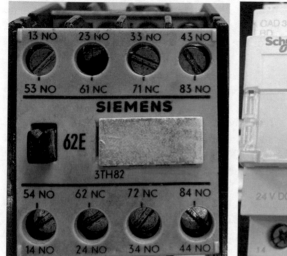 

**Fig. 11.23**  A real image of a magnetic contactor

**Fig. 11.24**  Numbering of a magnetic contactor

## 11.13  Fluid Power Logic Gates

An elementary block that has at least one input and one output is known as a logic gate. There are several logic gates normally used in fluid power. These are AND, OR, NOT and YES gates that are used as fundamental gates. The two inputs are used for all gates except NOT and YES gates.

### 11.13.1  AND Shuttle Valve

In fluid flow systems, the AND function is represented when several MPL (Moving part logic) valves are connected in series in a pipeline. In this type of connection, fluid flows exist as an output in the pipeline when both inputs are active (ON). Therefore, it is known as the dual pressure valve. The ISO and popular old symbols for pneumatic AND valves are shown in Fig. 11.25.

The mathematical expression for the output Z using the two inputs A and B is written as,

$$Z = A.B \tag{11.3}$$

The truth table for the AND logic gate is summarized in Table 11.1. Both inputs are pressurized as shown in Fig. 11.26a. Whereas input $A$ is pressurized and input $B$ is not pressurized as shown in Fig. 11.26b. From the truth table, it is observed that the output signal will be present at terminal $Z$ when the same compressed air signals are applied to the two inputs $A$ and $B$ at the same time as can be seen in Fig. 11.26a.

There will be no output signal if only one input signal is applied as shown in Fig. 11.26b. In the case of different pressure inputs, the lower pressure will reach the output. The logic gate symbol is shown in Fig. 11.27.

The AND logic gate and pneumatic symbols using 3/2 normally closed DCV and 2/2 normally closed DCV are shown in Fig. 11.27.

The AND function can be formed by connecting two input devices as push-button switches with output devices as relay coils. Here, the AND logic gate is formed using two start push-button switches are connected in series with the relay coil as shown in Fig. 11.28. In this circuit, the relay coil will be energized when the two push-button

**Fig. 11.25** A pneumatic AND valve

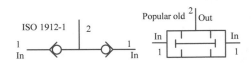

| Input A | Input B | Output Z |
|---------|---------|----------|
| 0 | 0 | 0 |
| 0 | 1 | 0 |
| 1 | 0 | 0 |
| 1 | 1 | 1 |

**Table 11.1** AND logic gate truth table

**Fig. 11.26** A logic AND gate

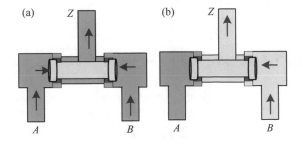

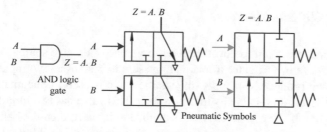

**Fig. 11.27**  A AND logic gate and pneumatic symbols

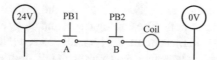

**Fig. 11.28**  Two push-button forms AND logic gate

switches press together. If one push-button is activated, then the circuit will not be completed as a result the coil will not be energized. The two 3/2 normally closed spring control push-button directional control valve is used to form a AND circuit that operates the single-acting cylinder as shown in Fig. 11.29a.

In this circuit, the two DCV's are pressed together to operate the single-acting cylinder. A dual-pressure valve is also used along with the two 3/2 normally closed spring control push-button directional control valves to operate the single-acting cylinder as shown in Fig. 11.29b. The red and the black DCVs need to be pressed at the same time to get the two inputs into the dual pressure valve. The cylinder will not operate if one input of the dual pressure valve is absent.

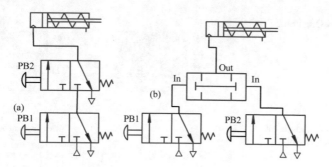

**Fig. 11.29**  Pneumatic symbols for AND logic gate

## 11.13.2   OR Shuttle Valve

A valve that allows fluid to flow through it from one of the two sources is known as the shuttle valve or OR gate. If unequal pressure exists at the two inputs ($A$ or $B$), an output will produce. Generally, the shuttle valve is used in pneumatic systems. Occasionally, it is used in the hydraulic system. This valve has two inputs ($A$, $B$) and one output ($Z$). If the compressed air is applied to input $A$, then the ball blocks the other input $B$ and the air will flow from $A$ to $Z$ as shown in Fig. 11.30a.

Similarly, when the compressed air is applied to the input $B$, then the ball blocks the other input $A$ and the air will flow from $B$ to $Z$ as shown in Fig. 11.30b. But, if the compressed air is connected to the two inputs then the airflow will reach the output from either of the inputs or both. The OR logic gate and the pneumatic logic symbols are shown in Fig. 11.31. In this circuit, 2/2 normally closed (NC) and normally open (NO) directional control valves are used to form an OR shuttle valve.

The parallel input devices such as push-button switches along with an output device form an OR function as shown in Fig. 11.32. Here, the relay coil is energized if either push-button switch is pressed. In the second circuit of Fig. 11.32, start and stop push-button switches are also used to form an OR function. In this case, the coil will be energized directly through the stop push-button switch. The coil will also be energized if pressing the start push-button switch.

## 11.14   Flow Amplification

A high-capacity cylinder requires a larger flow of compressed air. This large-capacity air can be dangerous to the relevant users. It is also unsafe to operate pneumatic

**Fig. 11.30**  A OR shuttle valve

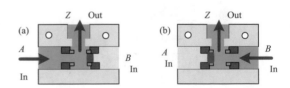

**Fig. 11.31**  A OR logic gate and pneumatic symbols

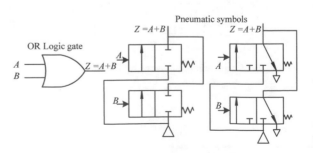

**Fig. 11.32** A OR logic
functions

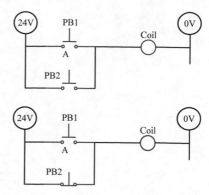

**Fig. 11.33** A flow
amplification circuit

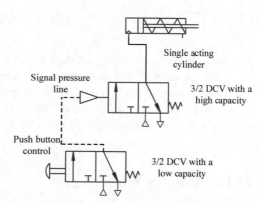

directional valves manually with a high flow capacity. Instead, a small directional control valve operates first and later, this can be used to operate the pneumatic control system with a large flow capacity. This type of pneumatic circuit is called flow amplification. The flow amplification usually ensures the safety of the operators. In a practical pneumatic circuit, the valves with a large flow capacity must be placed near the cylinder, while the valves with a smaller flow capacity should be placed near the control boards by providing a standard distance as shown in Fig. 11.33. When pressing the push button of 3/2 DCV, then the output will energize the external pilot of the 3/2 high capacity DCV and build up the pressure.

When the pressure of 3/2 pilot-operated DCV is higher than the spring pressure, then, it will provide a high capacity air to move the single-acting cylinder.

## 11.15 Signal Inversion

The signal inversion circuit produces the opposite output. The signal inversion circuit is shown in Fig. 11.34. Here, the 3/2 pilot-operated directional control valve will

**Fig. 11.34** A signal
inversion circuit

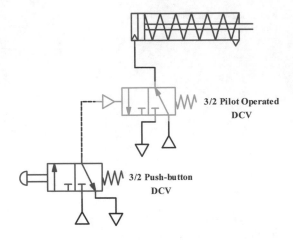

3/2 Pilot Operated
DCV

3/2 Push-button
DCV

produce output when the 3/2 push-button directional control valve is not operating.
The push-button DCV will produce output when pressing, then the 3/2 pilot-operated
directional control valve will produce no output. As a result, the cylinder will come
to its initial retraction position. Therefore, it is seen that at any given time, the output
of these two control valves is opposite to each other.

## 11.16   Memory Function

The memory circuit keeps the component at a certain position for some time until
there is a change of the signals. When pressing DCV 1, the output signal of DCV
3 will move the cylinder to its extended position. The cylinder will remain in this
position until pressing the DCV 2. When pressing DCV 2, it will produce another
signal that will move the cylinder to its retraction position. The circuit for memory
function is shown in Fig. 11.35

**Fig. 11.35** A memory
function circuit

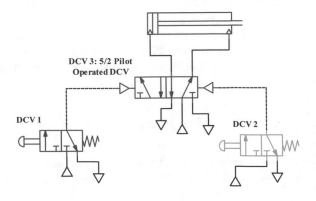

DCV 3: 5/2 Pilot
Operated DCV

DCV 1

DCV 2

## 11.17  Time Delay Valve

The time delay valve is used between the two operations of a pneumatic circuit. The time delay valve consists of the chamber (accumulator), spool, one-way flow control valve and connection ports. The output port A is connected to a single-acting cylinder. When a pressurized air enters into the inlet port, the air passes through a one-way flow control valve (consists of a basic check valve and adjustable orifice) and fills the accumulator. Then the pressure exerted on the spool valve makes the connection between the pressure port P to the output port A as shown in Fig. 11.36 and the cylinder extends.

The chamber (accumulator) will release pressure through a one-way flow control valve to an atmosphere. As a result, the air pressure at the chamber will be reduced slowly as shown in Fig. 11.37. Finally, when the air pressure of the chamber is less than the spring pressure, then the cylinder will move to its original or retraction position. The delay function is divided into ON-signal delay and OFF-signal delay. The ON-signal delay circuit is shown in Fig. 11.38.

This circuit consists of the 3/2 NC spring control DCV, flow control valve, accumulator, 5/2 external pilot DCV. Here, the one-way control valve acts as a resistor (restrictor) and the accumulator acts as a capacitor. When the push-button of a 3/2 NC DCV is pressed, it will allow airflow from the source to the flow control valve. This flow control valve slows down the flow of air to charge the accumulator. After fully charging the accumulator, the air pressure will enter the 5/2 DCV and build up the pressure. When this pressure is higher than the spring pressure, then 5/2 DCV will activate and change the state. Finally, the output signal of this valve will extend the cylinder. After the extension, a few seconds later the air will pass to the atmosphere through the flow control valve and 3/2 NC DCV. As a result, the pressure of the

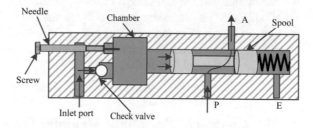

**Fig. 11.36**  A time delay valve for the extension

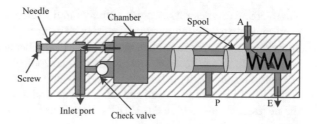

**Fig. 11.37**  A time delay valve for the retraction

**Fig. 11.38** A ON-signal delay circuit

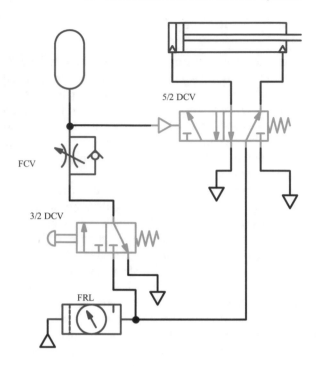

accumulator will reduce slowly and the cylinder will retract when the accumulator pressure is less than the spring pressure.

The OFF-signal delay circuit consists of the same components except the flow control valve needs to swap as shown in Fig. 11.39.

When the push-button of the 3/2 DCV is pressed, then air will flow freely through the flow control valve and charge the accumulator faster. The pressure of the 5/2 DCV will build up immediately and will change the state. As a result, the cylinder will extend. After a few seconds (longer than ON-signal delay) later, the air will pass slowly to the atmosphere through the flow control valve and 3/2 DCV. The cylinder will retract to its original position.

## 11.18   Control of Pneumatic Cylinder Option I

There are two types of solenoids namely single solenoid and double solenoids are used to control pneumatic cylinders. The cylinder will extend when the solenoid is energized and will retract when the supply is disconnected. A double-acting cylinder is controlled by a 5/2 single solenoid directional control valve as shown in Fig. 11.40. A control circuit with push-button switches (start and stop), magnetic contactor coil and solenoid is used to extend and retract the cylinder. A general ladder circuit along the necessary rungs, 24 and 0 V DC supply is used to operate the control circuit.

**Fig. 11.39** A OFF-signal
delay circuit

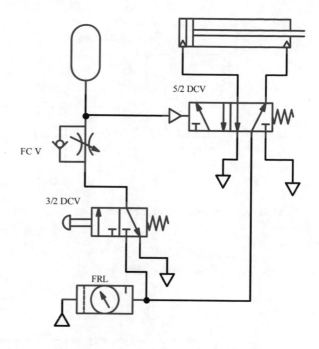

**Fig. 11.40** A general ladder
circuit

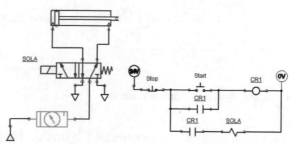

When the start push-button switch is pressed, then the current will flow through this line or rung that energizes the coil. Therefore, the open contacts of the magnetic contactor will be closed and the coil will remain in a closed position through the CR1-1 contact path unless to press the stop push-button switch.

As a result, the solenoid (SOLA) will be energized pushes the cylinder to an extended position. The control circuit along with the extended position of the cylinder is shown in Fig. 11.41. When the stop push-button switch is pressed, the magnetic contactor will be disconnected from the supply and de-energized. As a result, the solenoid SOLA will also be de-energized which in turn the cylinder will be retracted as shown in Fig. 11.42.

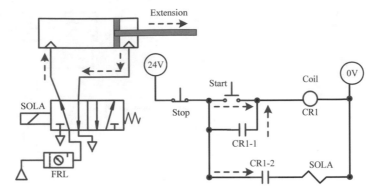

**Fig. 11.41**  A ladder circuit for an extension

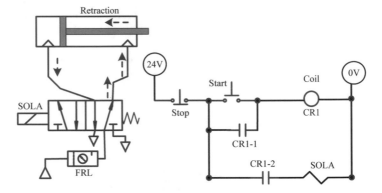

**Fig. 11.42**  A ladder circuit for retraction

## 11.19   Control of Pneumatic Cylinder Option II

A limit switch and a 5/2 single solenoid spring control directional control valve along with other necessary components are used to control a double-acting cylinder as shown in Fig. 11.43. The limit switch is placed at the starting position of the piston of the cylinder and this circuit is drawn using Automation Studio software [3]. The control circuit is drawn using a start push-button switch, magnetic contactor coil, normally open contact of the limit switch, and solenoid to extend and retract the cylinder. The 24 V DC supply is used to operate the control circuit.

Initially, the limit switch is an open condition as shown in Fig. 11.44. In an Automation Studio software, click on normal simulation then the limit switch will provide the closed path. Then press the start push-button switch and the cylinder will start to an extended position as shown in Fig. 11.44. After fully extend of the cylinder, the limit switch will provide an open connection that disconnects the specific rung

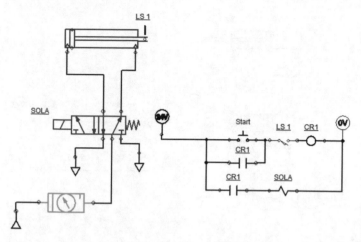

**Fig. 11.43** A pneumatic and electrical control for extension and retraction

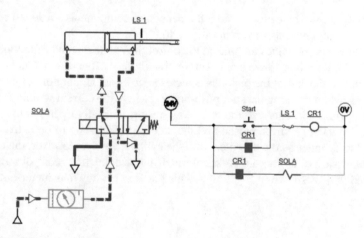

**Fig. 11.44** Circuit under normal and extended conditions

and de-energizes the coil. As a result, the cylinder will be retracted to its original position as shown in Fig. 11.45.

## 11.20 Control of Pneumatic Cylinder Option III

In this circuit, the limit switch is placed to the highest extended position of the piston of the cylinder. In Automation Studio software, it can be done by double-clicking on the cylinder and changing the extension value to 100%. Once it is done, then change the cylinder piston position to 0%. The limit switch, 5/2 single solenoid spring return

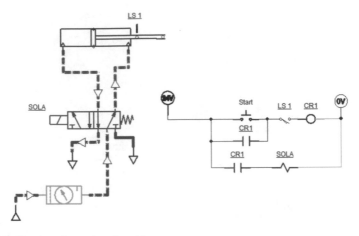

**Fig. 11.45** Circuit under retracted position

directional control valve along with other necessary components is used to control a double-acting cylinder as shown in Fig. 11.46.

Two flow control valves are placed to control the extension and retraction of the cylinder. This circuit is drawn using Automation Studio software and simulated. The limit switch at the end of the piston is at a closed position. Under normal simulation conditions, when pressing the start push-button switch, the current will flow through the limit switch and energize the coil. As a result, normally open contacts of the magnetic contactor will be closed and the solenoid (SOLA) will be activated that extends the cylinder as the air enters into the cylinder freely. However, the cylinder will extend slowly as the air passes through the second flow control valve in a controlled way as shown in Fig. 11.47. This position is known as meter-out during extension.

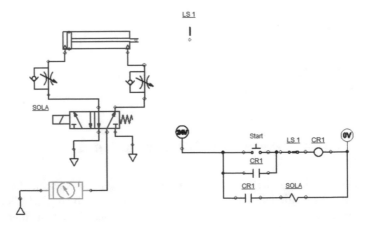

**Fig. 11.46** Circuit drawn Automation Studio software

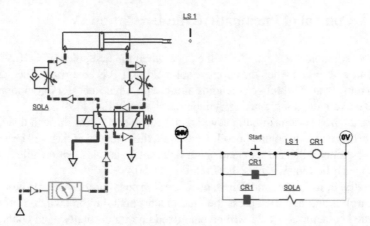

**Fig. 11.47** Cylinder under extension during a simulation

After the extension, when the piston touches the limit switch, then disconnect the supply that de-energizes the coil. As a result, the cylinder will retract slowly as the air comes out from the first flow control in a controlled way. This position of the cylinder is known as meter-out during retraction as shown in Fig. 11.48.

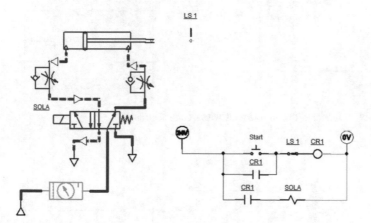

**Fig. 11.48** Cylinder under retraction during a simulation

## 11.21   Control of Pneumatic Cylinder Option IV

In double solenoids, the cylinder will extend when the first solenoid SOLA is energized and will retract when the opposite solenoid SOLB is energized. The 5/2 directional control valve with two solenoids along with other necessary components are
used to operate the double-acting cylinder as shown in Fig. 11.49.

Press the first start push-button switch, the current will flow through this line and
energize the magnetic contactor coil. As a result, the solenoid SOLA will be activated
and the cylinder will start to extend and it will remain at this position unless to press
the stop push-button switch as shown in Fig. 11.50.

For retraction of the cylinder, first, press the stop push-button switch to de-energize
the control circuit and then press the second start push-button switch. In this case,
magnetic contactor coil CR2 will be energized and its normally open contacts will

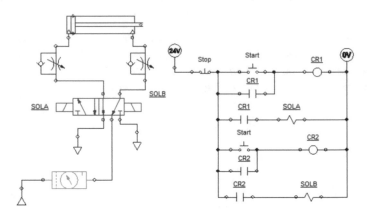

**Fig. 11.49**  Cylinder with two solenoids with a control circuit

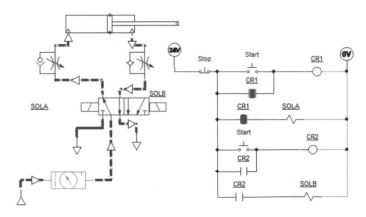

**Fig. 11.50**  Cylinder at extended position

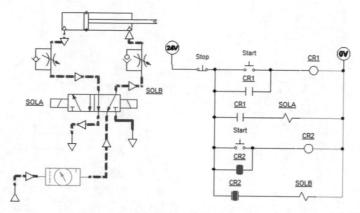

**Fig. 11.51** Cylinder at retraction position

be closed that will energize the solenoid SOLB which in turn the cylinder will be retracted to its original as shown in Fig. 11.51. These types of operation of the cylinder can be compared with the forward-reverse operation of a three-phase induction motor.

## 11.22   Control of Pneumatic Cylinder Option V

The 5/2 directional control valve with two solenoids, limit switches along with other necessary components are used to operate the double-acting cylinder as shown in Fig. 11.52.

In an electrical control circuit, when pressing the start push-button switch the current will flow and energize the magnetic contactor coil CR1 which in turn closes its normally open contacts and open its normally closed contacts. The current will

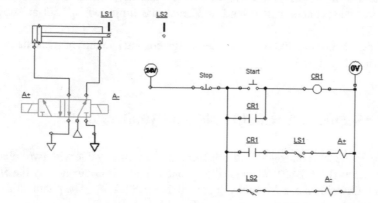

**Fig. 11.52**  Pneumatic circuit and a normal control circuit

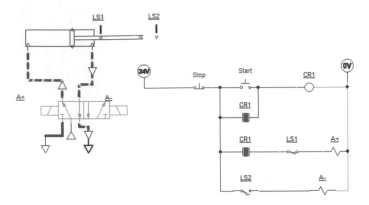

**Fig. 11.53**  Cylinder extended and nearly touches the limit switch LS2

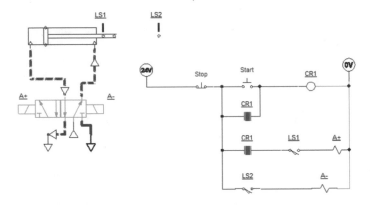

**Fig. 11.54**  Cylinder retracts as solenoid A- de-energizes

continue to flow through these closed contacts and keep energizing the coil CR1. Therefore, the cylinder will extend and touch the limit switch LS2 as shown in Fig. 11.53.

As a result, the solenoid A- will be de-energized which in turn retracts the cylinder as shown in Fig. 11.54.

## 11.23   Sequentially Control of Two Cylinders

The first cylinder is controlled by the 5/2 directional control valve with two solenoids, limit switches (LS1, LS2). Whereas the second cylinder is controlled by the 5/2 two solenoids directional control valve, limit switches (LS3, LS4). The pneumatic circuit and electrical control circuit is shown in Fig. 11.55.

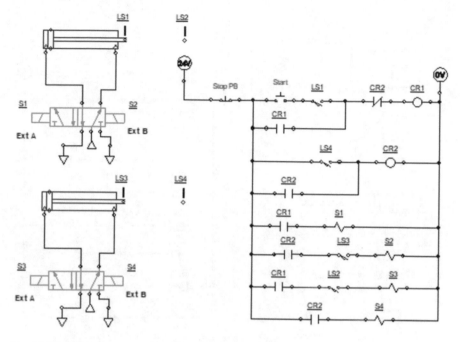

**Fig. 11.55** Pneumatic and an electrical control circuit for two cylinders

When pressing the start push-button switch, the current will flow and energize the magnetic contactor coil CR1 which in turn all associated open contacts of CR1 will be closed. As a result, the solenoid S1 will be activated first and extend the piston of cylinder 1. Then the solenoid S3 will be energized and the piston of cylinder 2 will extend as shown in Fig. 11.56.

Immediately after extension, the contactor coil CR2 will be de-energized which will also de-energize the limit switch LS4 and open it. As a result, the piston of cylinder 2 will retract. Then the magnetic contactor coil CR1 will be de-energized through the limit switch LS1 which in turn de-energizes the solenoid S2. Finally, the first cylinder will be retracted as shown in Fig. 11.57. Press the stop push-button switch to start the second cycle operation.

## 11.24  Basics of PLC

PLC is abbreviated as the programmable logic controller and it is a microprocessor-based solid device that works on the principle of logic gates. In the 1960s, the commercial PLC was designed and developed by Modicon to replace the relay for General Motors. In the 1970s, the microprocessor also played an important role to modernize the PLC. However, along with PLC, the relay is still used in the different

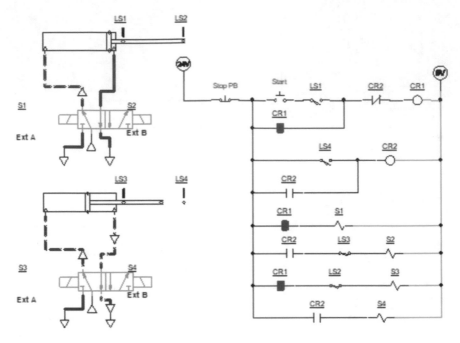

**Fig. 11.56** Cylinders extended due to energizing S1 and S3

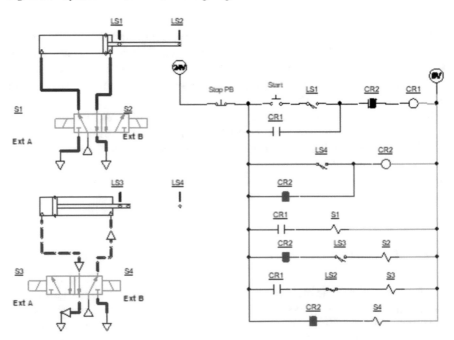

**Fig. 11.57** Cylinders retracted due to de-energizing S4 and S2

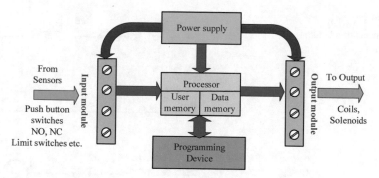

**Fig. 11.58** Common parts of PLC

control systems. A PLC is an industrialized personal computer that is used to control different types of industrial processes. The PLC reads system inputs, processes user-written software logic and switches outputs on or off positions.

The main components of PLC are inputs, outputs, memory, CPU and power supply as shown in Fig. 11.58. The inputs are classified as digital or binary status elements, high speed (same as binary but high frequency), analog (continuous value). Similarly, the outputs are classified as digital, analog and high speed. Whereas the memory is classified as RAM, ROM, EPROM etc. The processor will check the status of the inputs and give INPUTS to the input module or input image memory.

The inputs are push-button switches, NO and NC contacts, pressure switches, limit switches etc. Later, the CPU or processor will verify it with the logic and update the output to the output module or output image memory. Whereas the outputs are the lamp, relay coil, magnetic contactor coil, timer coil etc.

## 11.25 PLC Languages

Based on IEC 61,131–3 standard, the PLC language is classified as the following.

- Ladder Logic-It is the most common programming language for PLC.
- Mnemonic Instruction (MI)-It is also known as a statement list.
- Functional Block Diagram (FBD).
- Sequential Function Charts (SFC).
- Structured Text (ST)

Electrical Engineers usually use ladder logic, Electronic Engineers use FBD and other professionals use SFC, MI and ST.

**Fig. 11.59** Circuit components of a ladder diagram

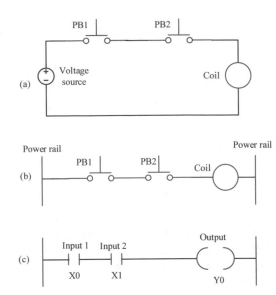

## 11.25.1   Ladder Programming

The most common form of language used in a PLC is the ladder programming language. The ladder diagram consists of two vertical lines that represent the power rails. The rung is the horizontal line that is connected between the power rails. Each program task in a ladder circuit is specified as through a rung. The rung is connected horizontally with two power rails. Consider a coil of the relay is connected in series with two push-button switches across a power source as shown in Fig. 11.59a. This coil will be energized when pressing the push-button switches together. This circuit can be redrawn by using a ladder diagram as shown in Figs. 11.59b and c.

## 11.25.2   Mnemonic Instructions

Some PLC manufacturers use Boolean language to control to program controller. This Boolean language of PLC is known as Boolean mnemonic language. A statement list is another way to represent the set of instructions of PLC. These sets of instructions such as input and output components are standardized by the international organization IEC6-1131-3. These sets of instructions in the statement list perform the same task as the ladder logic do. There are several companies mainly designing and providing PLC products. The mnemonic codes that are used by different companies are discussed in Table 11.2.

**Table 11.2**  Comparison of different mnemonic codes for different manufacturers

| IEC1131-3 | Mitsubishi | Omron | Simens | Ladder diagram | Operation |
|---|---|---|---|---|---|
| LD | LD | LD | A | Start a rung with an open contact | Load operand into result register |
| LDN | LDI | LD NOT | AN | Start a rung with a closed contact | Load negative operand into result register |
| AND | AND | AND | A | A series element with open contacts | Boolean AND |
| ANDN | ANI | AND NOT | AN | A series element with closed contacts | Boolean AND with a negative operand |
| OR | OR | OR | O | A parallel element with open contacts | Boolean OR |
| ORN | ORI | OR NOT | ON | A parallel element with closed contacts | Boolean OR with a negative operand |
| ST | OUT | OUT | = | An output from a rung | Store result register into the operand |

## 11.26  Logic Functions Representations

The main concepts of fundamental logic functions or gates are used in a ladder logic language. These are AND, OR, and NOT gates. Consider the two inputs A and B are normally open contacts connected in series (AND) to an out of a relay coil or lamp as shown in Fig. 11.60a.

In Fig. 11.60b, two inputs A and B are connected in parallel and then connected to the output lamp which forms an OR logic function. A normally closed contact is connected in series with an output lamp which forms a NOT logic function as shown in Fig. 11.60c. The XOR function can be formed by considering two normally closed inputs are connected in parallel and then connected with an output lamp as shown in Fig. 11.61. Here, two inputs are inverted as NOT logic gates and connected to OR gate as inputs as shown in Fig. 11.61b and its equivalent circuit is shown in Fig. 11.61c.

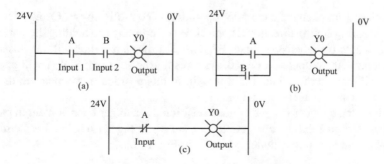

**Fig. 11.60**  Logic functions represented as ladder logic

**Fig. 11.61** Inverted inputs represented as ladder logic

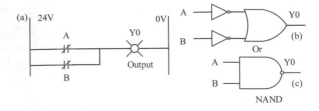

## 11.27   PLC Symbols and Cards

An input has two options either namely normally closed (NC) or normally open (NO) in a relay contact. Usually, the processor reads or scans the input/output (I/O). If it detects a device that is wired normally closed or normally open contact, then it will instruct PLC to change the state of the contact in the program. PLC manufacturer, Alen Bradley represents these types of activities as EXAMINE ON and EXAMINE OFF. The normally open contact is represented as EXAMINE ON or XIC Examine if closed. Whereas, the normally closed contact is represented as EXAMINE OFF or XIO Examine if open. In XIC contact, the processor will examine the reference address of this input and close the normally open contact to provide the power flow to the output. In XIO contact, the processor will examine the reference address of this input and keep the contact remains closed or OFF position if the program needs to operate. The XIO has a high bit-level address that is examined for an OFF (0/Low) condition.

The XIO and XIC may be a switch or pushbutton that are usually connected to the output. The output element OTE is abbreviated as output energize. The OTE means energized which means our output portion of our rung is turned on. The input and output cards provide the address for inputs and outputs, respectively. The XIC, XIO, OTE and PLC cards are shown in Fig. 11.62.

## 11.28   Control of Cylinder in PLC

A double-acting pneumatic cylinder operation is controlled by PLC and 5/2 single solenoid spring control directional control. When pressing and holding the start push-button switch, then the NO contact E3_1.IN0 of the start switch will be closed that will energize the output coil OUT0. As a result, the solenoid, SOLA will energize which will extend the cylinder and remain in this position until releasing the start push-button switch as shown in Fig. 11.63.

To hold the cylinder in the extended position use another open contact in parallel with the start push-button open contact as shown in Fig. 11.64. The PLC circuit to control a double-acting cylinder is shown in Fig. 11.65.

**Fig. 11.62**  PLC cards,
inputs and output

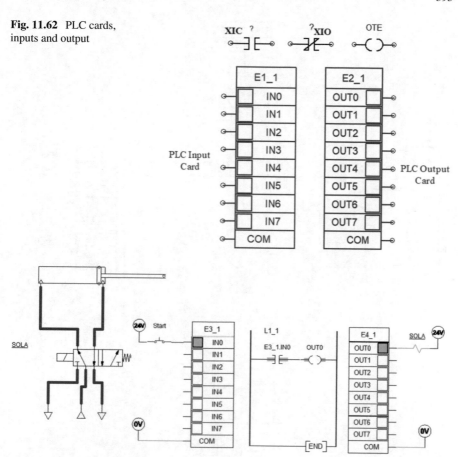

**Fig. 11.63**   Control of cylinder circuit 1

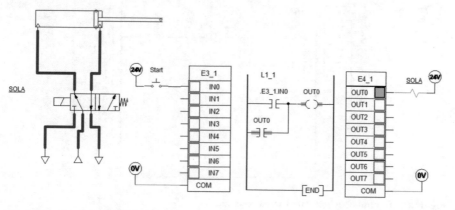

**Fig. 11.64**   Control of cylinder circuit 2

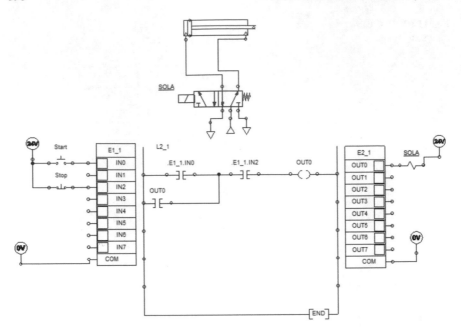

**Fig. 11.65** Control of cylinder circuit 3

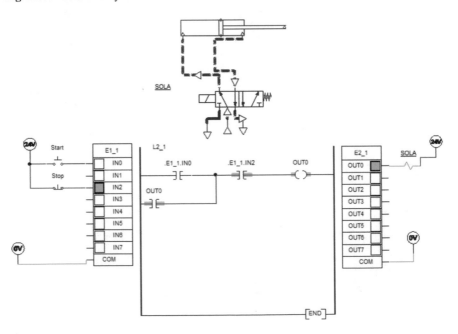

**Fig. 11.66** Control of cylinder for extension circuit 4

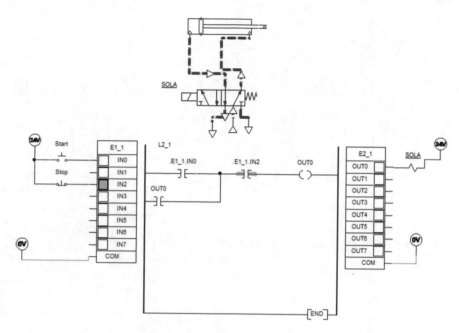

**Fig. 11.67** Control of cylinder for retraction circuit 5

The cylinder is extending when pressing the start push-button switch and remains in this position until to press the stop push-button switch as shown in Fig. 11.66. The retracting position of the cylinder is shown in Fig. 11.67.

The control of a double-acting cylinder with a proximity switch. The normally open contact of a proximity switch is connected to IN2 of the input card. When pressing the start push-button switch, then the cylinder will be extended as shown in Fig. 11.68. A timer ON delay (TON) is used to disconnect the circuit once preset value 3 s is reached and the cylinder will be retracted as shown in Fig. 11.69.

Another way to control a double-acting using limit switches. Here, the cylinder will continuously extend and retract until pressing the stop push-button switch. The extension and retraction circuits are shown in Figs. 11.70 and 11.71, respectively.

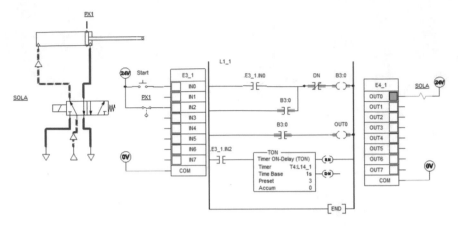

**Fig. 11.68**   Control of cylinder for extension circuit 6

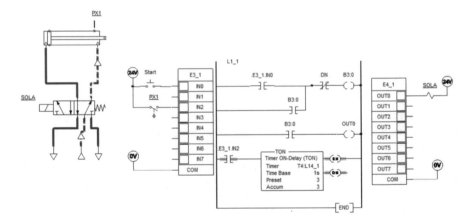

**Fig. 11.69**   Control of cylinder for retraction circuit 7

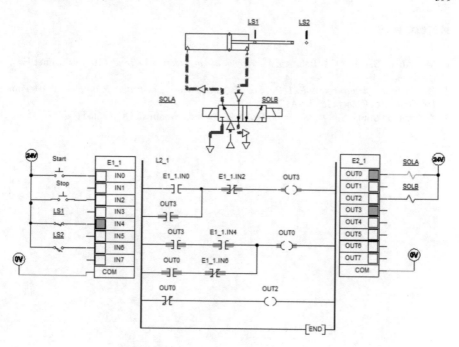

**Fig. 11.70** Control of cylinder for extension circuit 8

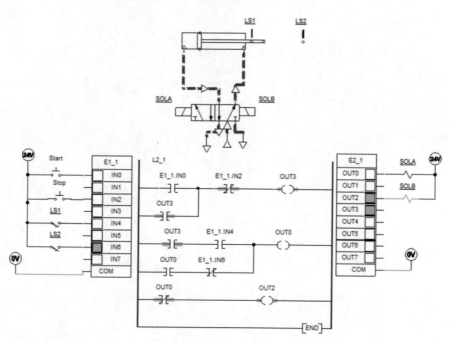

**Fig. 11.71** Control of cylinder for retraction circuit 9

# References

1. Md. Abdus Salam, Q.M. Rahman, *Fundamentals of Electrical Circuits Analysis*, First Edn. (Springer, 2018)
2. B.S. Elliott, *Electromechanical Devices & Components Illustrated Sourcebook*, 1st edn. (McGraw-Hill Companies, USA, 2007)
3. Automation Studio Software, (Femic Technologies Inc., Montreal, QC, Canada, 2020)

# Index

Printed in the United States
by Baker & Taylor Publisher Services